Wide Bandgap Semiconductor Materials and Devices 17

Editors:

J. M. Zavada

V. Chakrapani

S. Jang

T. J. Anderson

J. K. Hite

Sponsoring Division:

 Electronics and Photonics

Published by
The Electrochemical Society

65 South Main Street, Building D
Pennington, NJ 08534-2839, USA

tel   609 737 1902
fax   609 737 2743

www.electrochem.org

ecstransactions ™

Vol. 72, No. 5

Copyright 2016 by The Electrochemical Society.
All rights reserved.

This book has been registered with Copyright Clearance Center.
For further information, please contact the Copyright Clearance Center,
Salem, Massachusetts.

Published by:

The Electrochemical Society
65 South Main Street
Pennington, New Jersey 08534-2839, USA

Telephone 609.737.1902
Fax 609.737.2743
e-mail: ecs@electrochem.org
Web: www.electrochem.org

ISSN 1938-6737 (online)
ISSN 1938-5862 (print)
ISSN 2151-2051 (cd-rom)

ISBN 978-1-62332-357-8 (CD-ROM)
ISBN 978-1-60768-715-3 (PDF)

Printed in the United States of America.

Preface

The papers included in this issue of *ECS Transactions* were originally presented in the symposium "Wide Bandgap Semiconductor Materials and Devices 17", held during the 229th meeting of The Electrochemical Society, in San Diego, CA, from May 29 to June 2, 2016.

ECS Transactions, **Volume 72, Issue 5**

Wide Bandgap Semiconductor Materials and Devices 17

Table of Contents

Preface   *iii*

Chapter 1
Intersection of III-V and IV

(Invited) Electrothermal Performance Optimization of III-Nitride HEMTs Capped   3
with Nanocrystalline Diamond
   M. J. Tadjer, T. J. Anderson, T. I. Feygelson, K. D. Hobart, M. G. Ancona,
   A. D. Koehler, J. K. Hite, V. D. Wheeler, B. B. Pate, F. J. Kub, C. R. Eddy Jr.

Investigation of Nucleation and Intermixing at Hetero-Interface in III-Nitride-4H-SiC   9
Structures
   R. W. Enck, A. V. Sampath, R. Chung, D. B. Knorr Jr., G. A. Garrett, M. L. Reed

Chapter 2
Nitride Devices

GaN MIS-HEMT with Low Dynamic ON-Resistance Using SiON Passivation   19
   S. C. Liu, C. K. Huang, E. Y. Chang

(Invited) Improved GaN Based Hydrogen Sensors   23
   K. H. Baik, J. Kim, S. Jang

Chapter 3
Nitride Optoelectronics

On the Origin of the 4.7 eV Absorption and 2.8 eV Emission Bands in Bulk AlN   31
Substrates
   D. Alden, Z. Bryan, B. Gaddy, I. Bryan, G. Callsen, A. Koukitu, Y. Kumagai,
   A. Hoffmann, D. Irving, Z. Sitar, R. Collazo

v

Growth of GaN/InGaN Films and Heterostructures Via Super-Atmospheric MOCVD   41
  J. R. Krause, E. B. Stokes

(Invited) New Directions in GaN Photonics Enabled by Electrochemical Processes   47
  C. Zhang, G. Yuan, K. Xiong, S. H. Park, J. Han

Chapter 4
Oxides

Channel Scaling Behavior of Amorphous In-Zn-O Thin Film Transistors with High   59
Mobility over 35 cm^2/Vsec
  S. Lee, D. C. Paine

Investigation on Alumina Passivation for Improved IGZO TFT Performance   67
  T. Mudgal, N. Edwards, P. Ganesh, A. Bharadwaj, R. G. Manley, K. D. Hirschman

Ultra-Small ZnO Nanoparticles for Charge Storage in MOS-Memory Devices   73
  N. El-Atab, A. Nayfeh

Chapter 5
Poster Session

Thermal-Structural Optimization of Light with LED Packaging   83
  L. Zhang, Y. Li, D. Ge, X. D. Zhu

Mechanical Analysis of Stretchable AlGaN/GaN High Electron Mobility Transistors   89
  R. P. Tompkins, I. Mahaboob, S. Shahedipour-Sandvik, N. Lazarus

Effect of Ba Co-Doping on $Sr_5(PO_4)_3Cl:Ce^{3+}$ Blue Emitting Phosphor Application for   97
White Light Emitting Diodes
  G. Deressa

Author Index   109

Facts about ECS

The Electrochemical Society (ECS) is an international, nonprofit, scientific, educational organization founded for the advancement of the theory and practice of electrochemistry, electronics, and allied subjects. The Society was founded in Philadelphia in 1902 and incorporated in 1930. There are currently over 7,000 scientists and engineers from more than 70 countries who hold individual membership; the Society is also supported by more than 100 corporations through Corporate Memberships.

The technical activities of the Society are carried on by Divisions. Sections of the Society have been organized in a number of cities and regions. Major international meetings of the Society are held in the spring and fall of each year. At these meetings, the Divisions and Groups hold general sessions and sponsor symposia on specialized subjects.

The Society has an active publication program that includes the following:

Journal of The Electrochemical Society — (JES) is the leader in the field of electrochemical science and technology. This peer-reviewed journal publishes an average of 550 pages of 85 articles each month. Articles are published online as soon as possible after undergoing the peer-review process. The online version is considered the final version and is fully citable with articles assigned specific page numbers within specific issues. The date of online publication is the official publication date of record.

Journal of Solid State Science and Technology — (JSS) is one of the newest peer-reviewed journals from ECS launched in 2012. JSS covers fundamental and applied areas of solid state science and technology including experimental and theoretical aspects of the chemistry and physics of materials and devices. Articles are published online as soon as possible after undergoing the peer-review process. The online version is considered the final version and is fully citable with articles assigned specific page numbers within specific issues. The date of online publication is the official publication date of record.

Electrochemistry Letters — (EEL) is one of the newest journals from ECS launched in 2012. It is dedicated to the rapid dissemination of peer-reviewed and concise research reports in fundamental and applied areas of electrochemical science and technology. Articles are published online as soon as possible after undergoing the peer-review process. The online version is considered the final version and is fully citable with articles assigned specific page numbers within specific issues. The date of online publication is the official publication date of record.

Solid State Letters — *(SSL)* is one of the newest journals from ECS launched in 2012. It is dedicated to the rapid dissemination of peer-reviewed and concise research reports in fundamental and applied areas of solid state science and technology. Articles are published online as soon as possible after undergoing the peer-review process. The online version is considered the final version and is fully citable with articles assigned specific page numbers within specific issues. The date of online publication is the official publication date of record.

Electrochemical and Solid-State Letters — (ESL) was the first rapid-publication electronic journal dedicated to covering the leading edge of research and development in the field of solid-state and electrochemical science and technology. ESL was a joint publication of ECS and IEEE Electron Devices Society. Volume 1 began July 1998 and contained six issues, thereafter new volumes began with the January issue and contained 12 issues. The final issue of ESL was Volume 16, Number 6, 2012. Preserved as an archive, ESL has since been replaced by SSL and EEL.

Interface— *Interface* is an authoritative yet accessible publication for those in the field of solid-state and electrochemical science and technology. Published quarterly, this four-color magazine contains technical articles about the latest developments in the field, and presents news and information about and for members of ECS.

ECS Meeting Abstracts— *ECS Meeting Abstracts* contain extended abstracts of the technical papers presented at the ECS biannual meetings and ECS-sponsored meetings. This publication offers a first look into the current research in the field. ECS Meeting Abstracts are freely available to all visitors to the ECS Digital Library.

ECS Transactions— (ECST) is the online database containing full-text content of proceedings from ECS meetings and ECS-sponsored meetings. ECST is a high-quality venue for authors and an excellent resource for researchers. The papers appearing in ECST are reviewed to ensure that submissions meet generally-accepted scientific standards. Each meeting is represented by a volume and each symposium by an issue.

Monograph Volumes — The Society sponsors the publication of hardbound monograph volumes, which provide authoritative accounts of specific topics in electrochemistry, solid-state science, and related disciplines.

For more information on these and other Society activities, visit the ECS website:

www.electrochem.org

Chapter 1

Intersection of III-V and IV

2

Electrothermal Performance Optimization of III-Nitride HEMTs Capped with Nanocrystalline Diamond

M.J. Tadjer, T.J. Anderson, T.I. Feygelson, K.D. Hobart, M. Ancona, A.D. Koehler, J.K. Hite, V.D. Wheeler, B.B. Pate, F.J. Kub, C.R. Eddy, Jr

Naval Research Laboratory, 4555 Overlook Ave SW, Washington, DC 20375

AlGaN/GaN high electron mobility transistors (HEMT) capped with nanocrystalline diamond (NCD) have been demonstrated in the past to outperform electrically and thermally their SiN-passivated counterparts. However, a major process limitation for the integration of a diamond heat spreader has been the O_2-plasma damage in the gate opening associated with etching the diamond cap. A sacrificial gate (SG) process for plasma damage-free integration of top-side NCD capping layers is thus developed. On HEMTs with a SG, the addition of a NCD cap did not cause any significant degradation in mobility, carrier density, or sheet resistance. Hall characterization showed minimal (~6%) reduction in sheet carrier density and commensurate increase in sheet resistance, while maintaining mobility and on-state drain current density. Pulsed I_{DS} and on-resistance were improved, indicating that a 10 nm SiN/500 nm NCD could offer improved AlGaN surface passivation compared to a more conventional 100 nm thick PECVD SiN film.

Introduction

Gallium nitride (GaN) high electron mobility transistors (HEMT) are unique devices whose polarization charge confines a high density ($>10^{13}$ cm^{-2}) of electrons in a small volume quantum well near a heterostructure interface [1]. Of all compound semiconductor HEMT heterostructures, GaN-based HEMTs with an AlGaN barrier have offered the highest power density due to the wide bandgap and the high critical field of GaN and its ternary alloys. For these reasons, commercially available AlGaN/GaN HEMTs currently offer low-loss power and RF switching capabilities. Despite incredible progress, however, reliability concerns have persisted over the past two decades, particularly due to thermal limitations during simultaneous high field, high power operation of these devices. As a result, efficient thermal management of III-Nitride HEMTs is still an important issue that needs to be adequately addressed. A number of groups have approached the problem by integrating either a diamond cap or a diamond substrate in their fabrication process [2-5]. We have previously demonstrated that HEMTs capped with nanocrystalline diamond (NCD) perform not only thermally, but also electrically better than their SiN-passivated counterparts [6, 7]. However, a major process limitation for the top-side integration of a diamond heat spreading cap has been the O_2-plasma damage in the gate area opening associated with etching the diamond cap. To reduce plasma damage, Anderson et al. have demonstrated a two-step diamond etch process, where the power is reduced towards the end of the dry etch to minimize plasma-induced damage to the AlGaN surface [6]. While this process has enabled the demonstration of improved electro-thermal performance of NCD-capped HEMTs, optimizing passive heat spreading requires NCD films with thermal conductivity as high as possible. Since the thermal conductivity of NCD is directly related to film thickness, as grain size scales with thickness, NCD capping layers thicker than 0.5 μm are desirable [8]. As a result, a two-step NCD etch process will be more difficult to calibrate, and a new solution for a gate opening in the heat spreading NCD film would potentially be required. The sacrificial gate (SG) process proposed in this work addresses the challenge of protecting the AlGaN surface when thick films of NCD are deposited, and further provides a robust solution for processing complications such as on-wafer variation in NCD thickness [9].

Experimental Details

Samples of AlGaN/GaN on a 4H-SiC substrate were processed into HEMT devices. After an inductively-coupled Cl_2-plasma isolation etch, the Ti/Al/Ni/Au Ohmic stack was lifted off and annealed at 850 °C for 30 sec in N_2. The resulting mesa isolation current was about 17 nA/mm at 100 V, and the contact resistance measured by the TLM method was about 1.3 Ω/mm. Then the SG was processed. First, a 10 nm thick Al_2O_3 layer was grown by atomic layer deposition (ALD), followed by a 100 nm thick PECVD-grown SiN_x layer. The SiN_x layer served as an etch stop for the subsequent etch of the NCD layer, and the Al_2O_3 served as an etch stop for the SiN_x layer. The Al_2O_3/SiN_x stack was patterned with the nominal gate dimension plus an additional 1 μm on each side in order to eliminate aligner registration error during metal gate realignment. Thus, the SG length was 4 μm, whereas the actual metal gate length was 2 μm ($W_G = 75$ μm). Another 10 nm of blanket SiN_x was deposited, followed by a blanket 0.5 μm thick NCD conformal growth over the device by microwave plasma CVD at 750 °C for 5 hours. Details of CVD NCD growth and resulting thickness-dependent thermal conductivity have been published elsewhere [10-12].

The Ni/Au gate feature was then realigned and a high power O_2-plasma ICP process was used to clear the NCD film in the gate region. This process resulted in a much more reliable gate opening since the etch stop was on a SiN film and a >20% built-in NCD overetch was not detrimental to the AlGaN surface. Subsequently, the SG was etched away using a SF_6-plasma ICP process. Because the PECVD SiNx gate finger was exposed to the NCD growth conditions, a significant overetch in the SF_6 ICP process was implemented. Finally, the Al_2O_3 layer was removed in dilute HF solution, and a 20/200 nm Ni/Au gate metal stack was defined.

Electrical Characterization

To quantify the effect of the O_2-ICP diamond etch process on the 2DEG charge in the HEMT, Hall characteristics were measured on 4 HEMTs, depending on whether a diamond cap was present and whether a SG process was employed, as opposed to the standard mid-process NCD capping process described in previous reports [4, 7, 13]. The Hall characterization results are summarized in Table I. When the SG process was not employed (rows 1 and 2), the NCD-capped HEMT (row 2) experienced a 36% decrease in mobility and a 30% decrease in sheet carrier density, leading to a 123% increase in sheet resistance.

TABLE I. Hall characteristics of AlGaN/GaN HEMTs with/without a sacrificial gate and with/without a diamond cap. Process conditions (SiN deposition, NCD etch, etc.) were identical among the samples.

NCD	SG	R_{SH} ($\Omega/\square$)	μ_H (cm^2/V·s)	N_{SH} (x10^{12} cm^{-2})
N	N	462	1538	8.78
Y	N	1031	985	6.15
N	Y	329	2100	9.02
Y	Y	350	2120	8.4

On the other hand, etching the NCD cap on the SG NCD-HEMT (row 4) did not cause any significant degradation in sheet resistance and carrier density (ΔR_{SH} = 6.4%, ΔN_{SH} = -6.8%), and the Hall mobility remained essentially unchanged. The origin of this small degradation is currently being attributed to the SF$_6$-plasma etch of the SG finger. It is expected that this damage can be completely removed by optimization of the SiN$_x$/Al$_2$O$_3$ dielectric stack.

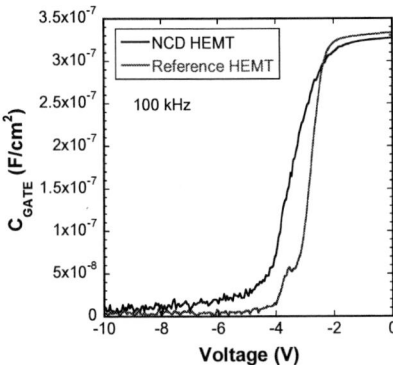

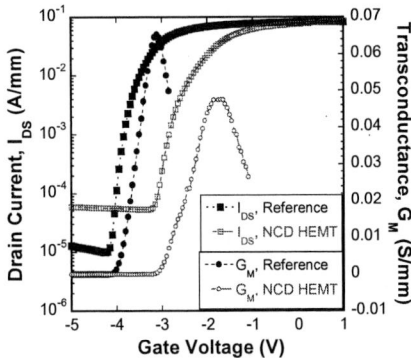

Figure 1. Capacitance-voltage curves measured on 150 µm diameter gate test structures at 100 kHz on SG-HEMTs with and without NCD capping layers.

Figure 2. Static (DC) I_{DS}-V_{GS} (V_{DS} = 1 V) characteristics of SG-HEMTs with and without NCD capping layers.

In the subthreshold region, the decrease in the slope of the C_G-V_G curve of the SG NCD-HEMT (Fig. 1) indicated a post SG-process increase in AlGaN surface trap density within the gate recess [14]. Furthermore, the off-state current for the SG NCD-HEMT increased by about a factor of 3 and the threshold voltage shifted by +1 V, as shown in Fig. 2, also consistent with fluorination of the AlGaN surface. Therefore, while the SG process achieved its purpose of maintaining the on-state transport characteristics of the 2DEG (e.g., maintain sheet charge), further optimization of the gate stack to maintain performance in the off-state is required.

In the on state (V_{GS} = +1 V), the drain-source current density for the SG NCD-HEMT was 93.5 mA/mm at V_{DS} = 1 V, comparable to that of the reference HEMT. The transconductance curve was stretched out over a wider bias range on the gate, with its $G_{M,MAX}$ value reduced, again as a consequence of the sacrificial gate finger overetch after exposure of the 100 nm PECVD SiN to the diamond growth process.

We note that only the AlGaN surface underneath the gate (L_G = 2 µm) was affected by this process limitation. In the access region (10 µm gate-drain, 2 µm gate-source spacings), the NCD cap was 10 nm away from the AlGaN surface, separated only by the thin SiN nucleation layer. A pulsed I-V characterization technique, sensitive to traps in the access region, was performed using an Accent DiVAD265EP instrument. Pulsed I_{DS}-V_{DS} ($V_{GS} = 0$ V) curves were measured before and after off-state quiescent stress. Figure 3a shows the improved current of the SG NCD-HEMT, suggesting either that traps at or near the HEMT surface were better passivated by using a 10 nm SiN / 500 nm NCD cap than by using a 100 nm thick SiN film (reference HEMT). Subsequently, the device was subjected to quiescent off-state drain stress ($V_{DS,Q} = 0$-50 V). The degradation in on resistance resulting from the 50 V pulsed stress (Fig. 3b), was about a factor of 2 lower for the NCD capped HEMT, compared to the reference device.

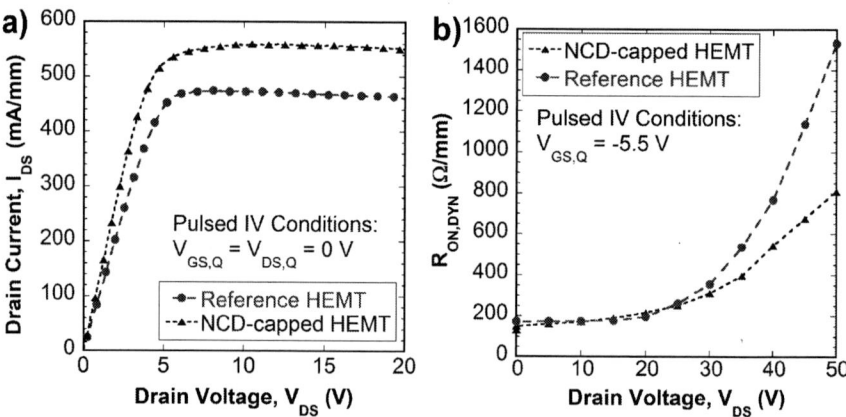

Figure 3. a) Pulsed I_{DS}-V_{DS} ($V_{GS} = 0$ V) characteristics of the NCD-capped HEMT with a SG process (SG NCD-HEMT), compared to those of a reference HEMT, b) Dynamic on resistance as a function of off-state drain bias stress.

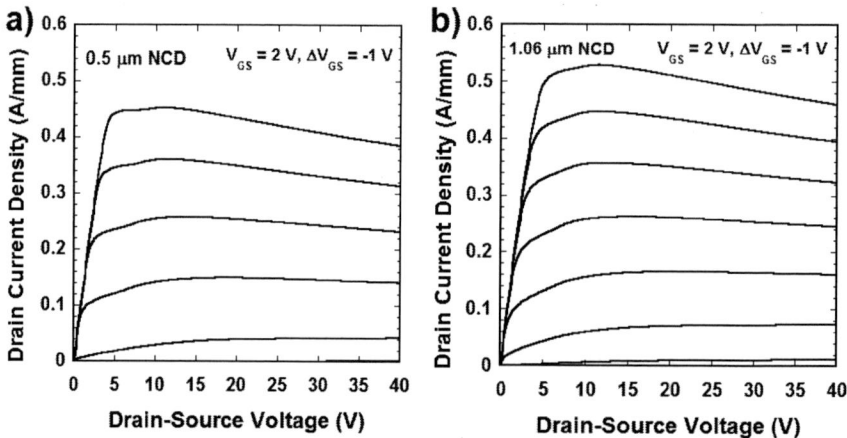

Figure 4. Static (DC) I_{DS}-V_{DS} characteristics of AlGaN/GaN HEMT capped with a) 0.5 μm and b) 1.06 μm NCD, respectively.

Static (DC) I_{DS}-V_{DS} output characteristics of SG NCD-HEMTs with 0.5 μm and 1.06 μm thick NCD capping layers are shown in Fig. 4. The sample with 1.06 μm thick NCD cap exhibited higher output current with a similar negative differential output resistance in the on state, indicating that an improved thermal profile was not the origin of the improved DC output power (18.4 W/mm). However, this result indicated that in order to improve electrical performance at high output power levels (> 10 W/mm), an NCD cap of greater than 0.5 μm thickness can be incorporated into the device.

Discussion

By means of a finite element thermal model, Wang et al. have demonstrated that current loss due to negative differential resistance induced by self-heating can be mitigated using a thicker, higher thermal conductivity diamond film [8]. We note that Wang's model had assumed a constant thermal conductivity for the diamond films. Philip et al., among other groups, have demonstrated that as the NCD film is grown thicker, its columnar structure would cause vertical grains of single crystal NCD to expand in width. Thus, limitations to thermal conductivity induced by grain boundaries will be less pronounced for thicker diamond films. For example, increasing diamond thickness from 0.5 to 1 μm will cause its thermal conductivity to double from approximately 4 W/m-K (about the same as that of Cu) to 8 W/m-K [10]. However, a thicker diamond cap would increase the aspect ratio of the gate recess and thus introduce additional processing challenges for high frequency diamond capped devices with scaled gate length. On the other hand, Anaya et al. have demonstrated that in-plane thermal conductivity of diamond for a given thickness can be improved by controlling the lateral evolution of diamond grains during growth [15]. Incorporation of high in-plane thermal conductivity diamond would thus enable sub-μm gate scaling in future process development.

The purpose of the SG process described in this work was to enable the integration of a thick NCD cap in a GaN HEMT device process that also requires preserving the integrity of the III-N surface and, thus, the performance of the 2DEG and device. It is therefore expected that simultaneously increasing the thickness and thermal conductivity of the diamond cap in future process iterations will introduce measurable improvements in HEMT output power without sacrificing device performance and reliability [16].

Conclusion

We showed that high power O_2-plasma did not degrade HEMT Hall parameters even when the diamond cap was significantly over-etched in the gate region. Static I-V characteristics were maintained in the on-state, while improved pulsed I-V response suggested improved passivation by 10 nm SiN/500 nm NCD compared to the standard 100 nm SiN. Maintaining off-state characteristics requires further process improvements such as post-SiN etch annealing, for instance. It is also likely that a thicker Al_2O_3 film would lead to improved post-SG process AlGaN surface, resulting in reduced gate trap density and improved G_M and $I_{DS,MAX}$ performance metrics.

Acknowledgments

The authors are sincerely grateful to Dr. Scott Sheppard (Wolfspeed) for providing device wafers, the NRL Nanoscience Institute support staff for support of processing equipment, and Mr. David Shahin for assistance with Keithley 4200 measurements. Research at NRL was supported by the Office of Naval Research.

References

1. O. Ambacher, B. Foutz, J. Smart, J.R. Shealy, N.G. Weimann, K. Chu, M. Murphy, A.J. Sierakowski, W.J. Schaff, L.F. Eastman, R. Dimitrov, A. Mitchell, and M. Stutzmann, J. Appl. Phys. 87, 334 (2000).
2. K. D. Chabak, J. K. Gillespie, V. Miller, A. Crespo, J. Roussos, M. Trejo, D. E. Walker, Jr., G. D. Via, G. H. Jessen, J. Wasserbauer, F. Faili, D. I. Babi, D. Francis, and F. Ejeckam, IEEE Electron Devices Lett., vol. 31, no. 2, pp. 99–101, Feb. 2010.
3. M. Alomari, M. Dipalo, S. Rossi, M.-A. Diforte-Poisson, S. Delage, J.-F. Carlin, N. Grandjean, C. Gaquiere, L. Toth, B. Pecz, and E. Kohn, Diamond Relat. Mater., vol. 20, no. 4, pp. 604–608, Apr. 2011.
4. M.J. Tadjer, T.J. Anderson, K.D. Hobart, T.I. Feygelson, J.D. Caldwell, C.R. Eddy, Jr., F.J. Kub, J.E. Butler, B.B. Pate, and J. Melngailis, IEEE Electr. Dev. Lett., vol. 33, no. 1, pp. 23-25, 2012.
5. K. Hirama, M. Kasu, and Y. Taniyasu, IEEE Electr. Dev. Lett., vol. 33, no. 4, pp. 513-515, 2012.
6. T.J. Anderson, K.D. Hobart, M.J. Tadjer, T.I. Feygelson, E.A. Imhoff, D.J. Meyer, D.S. Katzer, J.K. Hite, F.J. Kub, B.B. Pate, S.C. Binari, C.R. Eddy, Jr., 70th Dev. Research Conf. Proc., pp. 155-156, 2012.
7. D.J. Meyer, T.I. Feygelson, T.J. Anderson, J.A. Roussos, M.J. Tadjer, B.P. Downey, D.S. Katzer, B.B. Pate, M.G. Ancona, A.D. Koehler, K.D. Hobart, C.R. Eddy, Jr., IEEE Electr. Dev. Lett., vol. 35, no. 10, pp. 1013-1015, 2014.
8. A. Wang, M.J. Tadjer, and F. Calle, Semicond. Sci. Tech-nol., vol. 28, pp. 055010 (2013).
9. T.J. Anderson, A.D. Koehler, M.J. Tadjer, K.D. Hobart, T.I. Feygelson, J.K. Hite, B.B. Pate, C.R. Eddy, Jr., F.J. Kub, CS Mantech Conf. Digest, pp. 205-208, 2013.
10. J. Philip, P. Hess, T. Feygelson, J. E. Butler, S. Chattopadhyay, K. H. Chen, and L. C. Chen, J. Appl. Phys. 93, 2164 (2003).
11. J.E. Butler and A.V. Sumant, Chem. Vapor Deposition, vol. 14, no. 7/8, pp. 145-160, 2008.
12. E. Bozorg-Grayeli, A. Sood, M. Asheghi, V. Gambin, R. Sandhu, T.I. Feygelson, B.B. Pate, K.D. Hobart, and K.E. Goodson, "Thermal conduction inhomogeneity of nanocrystalline diamond films by dual-side thermoreflectance," Appl. Phys. Lett., vol. 102, no. 11, pp. 111907-1–111907-4, 2013.
13. A. Wang, M.J. Tadjer, T.J. Anderson, R. Baranyai, J.W. Pomeroy, T.I. Feygelson, K.D. Hobart, B.B. Pate, F. Calle, and M. Kuball, IEEE Trans. Electron Devices, vol. 60, no. 10, pp. 3149–3156, 2013.
14. D. Deen and J. Champlain, Appl. Phys. Lett. 99, 053501 (2011).
15. J. Anaya, S. Rossi, M. Alomari, E. Kohn, L. Tóth, B. Pécz, K.D. Hobart, T.J. Anderson, T.I. Feygelson, B.B. Pate, M. Kuball, "Control of the in-plane thermal conductivity of ultra-thin nanocrystalline diamond films through the grain and grain boundary properties," Acta Materialia 103 (2016) 141-152.
16. M. G. Ancona, S. C. Binari, and D. J. Meyer, J. Appl. Phys., vol. 111, no. 7, pp. 074504-1–074504-16, 2012.

Investigation of Nucleation and Intermixing at Hetero-interface in III-Nitride-4H-SiC Structures

R. W. Enck[a], A.V. Sampath[a], R.B. Chung[a], and D. Knorr[b], G. A. Garrett[a], M. L. Reed[a]

[a] U.S. Army Research Laboratory, RDRL-SEE-I, Adelphi, MD 20783
[b] U.S. Army Research Laboratory, RDRL-WMM-A, Aberdeen Proving Ground, MD, 21005

III-Nitride/SiC heterostructure devices are a promising approach for improving the performance of SiC devices such as bipolar transistors and avalanche photodiodes. However, the performance of these devices should critically depend on the properties of the hetero-interface that will likely lie within an active region of the device and impact carrier transport. Importantly, intermixing at the hetero-interface can effect doping profiles in these structures as constituent atoms of each semiconductor act as a dopant in the other. In this paper we explore the impact of *in situ* substrate preparation and migration enhanced epitaxy (MEE) on the nucleation and impurity concentration of thin AlN films grown by plasma-assisted molecular beam epitaxy on 4H-SiC. The surface morphology of the samples were examined by atomic force microscopy and the composition of the films were studies by secondary ion mass spectroscopy (SIMS) and depth profiling x-ray photoemission spectroscopy (XPS). The MEE approach promotes the nucleation of AlN growth on SiC in a 2D mode while suppressing the migration of Al into SiC. Active N that leaks around the closed shutter during in-situ preparation prevents the nucleation of AlN in a 2D growth mode at lower substrate temperatures that is attributed to GaN island formation at the hetero-interface.

Introduction

III-Nitride/SiC heterostructure devices are a promising approach for improving the performance of SiC devices such as bipolar transistors and avalanche photodiodes [1-4]. GaN/SiC heterojunction bi-polar junction transistors (HBT) can take advantage of the wider bandgap GaN base region to improve the current gain over SiC HBTs [2]. III-Nitride/SiC heterostructure avalanche photodiodes (APDs) are an exciting approach for extending the performance of 4H-SiC throughout the ultraviolet spectrum by either improving the absorption of near ultraviolet photons or mitigating the loss of carriers generated by deep ultraviolet photons to surface recombination over homogenous SiC APDs [3-4]. However, the critical requirement for efficient carrier transport at the hetero-interfaces of these devices requires optimization that considers the lattice mismatch between the III-Nitride semiconductor and 4H-SiC (between 1-4% for an $Al_xGa_{1-x}N$ alloy), impurities associated with the initiation of heteroepitaxial growth as well as intermixing as Si and C are a n-type dopant and deep-level center in GaN while Al and N are common p- and n-type dopants in 4H-SiC. Furthermore, the difference in spontaneous polarization between the III-Nitride semiconductor and 4H-SiC has been demonstrated to impacts carrier transport at the hetero-interface of III-Nitride/SiC heterojunction photodetectors due to the polarization induced interface charge [3-4]. Rodak et al. have shown that a large dipole barrier forms across the AlN region of a n-Al_xGa_{1-x}N/AlN/i-SiC/p-SiC diode that is sufficiently large to inhibit the near ultraviolet response of the diode and improve its solar-blind rejection [4]. The barrier is attributed to larger spontaneous polarization of AlN over both Al_xGa_{1-x}N and SiC and its height may be controlled by the thickness of the AlN barrier layer.

Okamura et al. have studied the impact of the presence of active N in the growth chamber during the Ga pre-deposition step used for the removal of surface oxides on the structural quality of AlN grown on 6H-SiC by plasma-assisted molecular beam epitaxy [5]. They observed that having the N plasma source started, but with the N shutter closed, during the Ga pre-deposition step resulted in a three-dimensional nucleation of AlN on SiC and an increase in threading and screw dislocations. In this paper, we explore the impact of migration enhanced epitaxy (MEE) as well as the presence of active N during the Ga pre-deposition step on the growth mode during nucleation of AlN on 4H-SiC by plasma-assisted molecular beam epitaxy (PA-MBE) as well as intermixing at the AlN/4H-SiC hetero-interface.

Design and Experiment

To explore the impact of in-situ Ga pre-deposition conditions, substrate growth temperature and MEE growth on the nucleation and impurity concentration of thin AlN films grown by plasma-assisted molecular beam, a series of 30nm thick AlN films were deposited on i-p epitaxial SiC structures grown on 4 degree miscut towards [11-20] 4H-SiC substrates. Prior to loading, all substrates were degreased for 20 minutes in acetone, methanol, and isopropyl alcohol. Then the substrates were cleaned using the following procedure: 1 minute rinse in deionized (DI) water, 10 minute soak in a heated solution of $NH_3O_4:H_2O_2:H_2O$ 1:1:5, a 1 minute rinse in distilled (DI) water, a 1 minute soak in $HF:H_2O$ 1:50, a 10 minute soak in a heated solution of $HCl:H_2O2:H_2O$, a 1 minute rinse in DI water, a 45 second soak in buffered oxide etch (BOE), and finally a 20 minute rinse in DI water. Next, the substrates were placed in the load-lock and outgassed at 250 °C for 12 hours in a vacuum pressure of less than 10^{-6} Torr. Subsequently, each substrate was loaded into the buffer chamber and outgassed at 600 °C for 12 hours in a vacuum pressure less than 10^{-8} Torr. All of the AlN films were deposited under N-limited conditions to promote smooth 2D film growth.

The Ga pre-deposition treatment was performed in the growth chamber at 740 °C and consisted of 10 repetitions of exposing the substrate surface to Ga for 2 mins at a deposition rate of YY followed by desorbing the Ga for 7 minutes. To understand the role of having active N present in the chamber, this was performed with the N plasma started and the shutter closed, or with it off until after the Ga pre-deposition was complete. After this step, the AlN film was grown at a substrate temperature of 600 °C or 740 °C. The AlN film was grown using either a direct growth approach without interruption, or the film was grown one monolayer at a time by alternating the Al and N shutters for 11 seconds, or the time required for one monolayer of each element to arrive on the surface. This technique is commonly referred to as MEE because it has been shown to improve surface roughness by enhancing adatom migration. Table 1 summarizes the growth conditions for the samples investigated.

The surface morphology of the AlN films were characterized by atomic force microscopy (AFM). The impurity concentrations in the films were examined by secondary ion mass spectroscopy by Evans Analytical Inc and/or depth profiling x-rays photo-emission spectroscopy (XPS). The XPS was performed on a Physical Electronics VersaProbeII instrument with an Al Kα source. Compositions were obtained from high resolution scans with a pass energy of 23.5 eV and a step size of 0.05 eV for C 1s, Al 2p, Si 2p, O 1s, N 1s, and Ga 3d regions. Depth profiles were obtained using an Ar ion gun operating at 2kV in intervals of 60 seconds. The high resolution scans mentioned above were repeated after each sputtering interval, and the data were processed using CasaXPS software.

TABLE I. Sample and results summary.

Sample	Plasma Prep	Growth Mode	Subs Temp (°C)	RMS Roughness (nm)	SIMS/XPS
A	Off	MEE	600	0.30	No mixing
B	Off	Direct	600	0.89	Al in SiC, Si in AlN
C	Off	MEE	740	1.2	Oxygen, no mixing
D	On	MEE	740	7.4	Ga, GaN present
E	On	MEE	600	6.9	No data

Results and Analysis

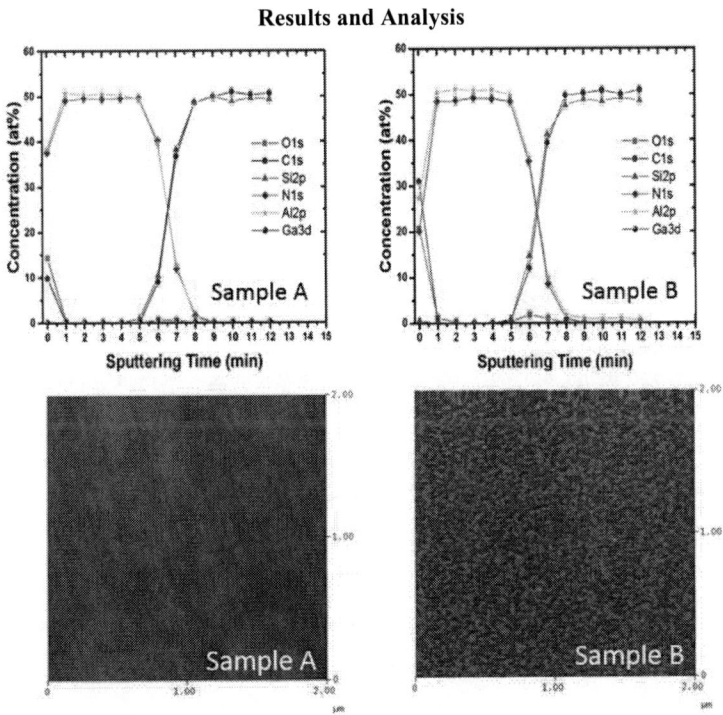

Figure 1. Atomic force microscopy and XPS signatures (inset) for migration enhanced (Left: A) and direct (Right: B) growth conditions.

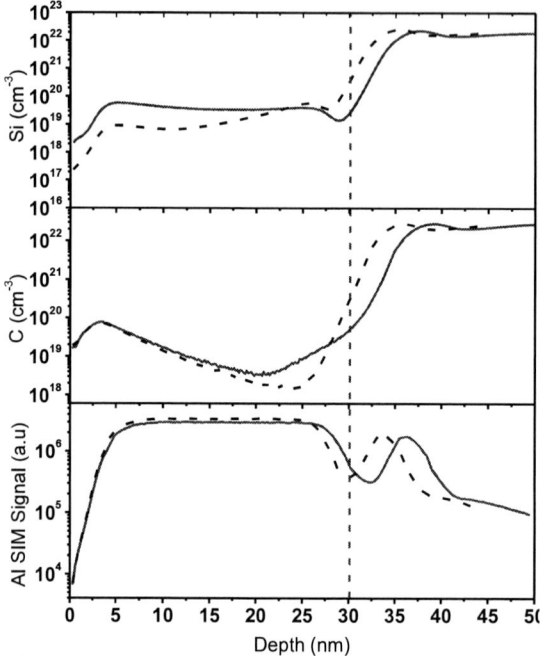

Figure 2. Secondary ion mass spectroscopy for migration enhanced (A) and direct (B) growth conditions.

Figure 1 compares the morphology and the XPS composition profile of the AlN film grown by MEE (sample A) and the direct grown (sample B) with the N plasma off during the Ga preparation. Sample A exhibits an atomically smooth, 2D surface with a root mean square roughness (R_{rms}) of 0.30 nm, and a nearly stoichiometric Al to N composition in the AlN layer that drops off at the hetero-interface. In contrast, sample B has a higher R_{rms} of 0.89 nm on a 2x2 μm scale with an absence of discernable steps on the substrate. XPS results indicate a slightly larger Al to N composition in the film (no visible droplets) for sample B and a detectable presence of Al in the SiC layer indicating migration. These results suggests that AlN film grown directly on SiC under these conditions nucleates three dimensionally in disparate places, leaving much of the SiC surface exposed to liquid metal Al during the start of growth. As a result, Al is likely able to migrate into SiC substitutionally for Si that has been reported by Hoke et al. to leave the SiC surface under these growth conditions [6]. In contrast, the MEE approach promotes the nucleation of AlN growth on SiC in a 2D mode that results in the presence of visible steps, better stoichiometry in the film as well as suppresses the diffusion of Al into SiC.

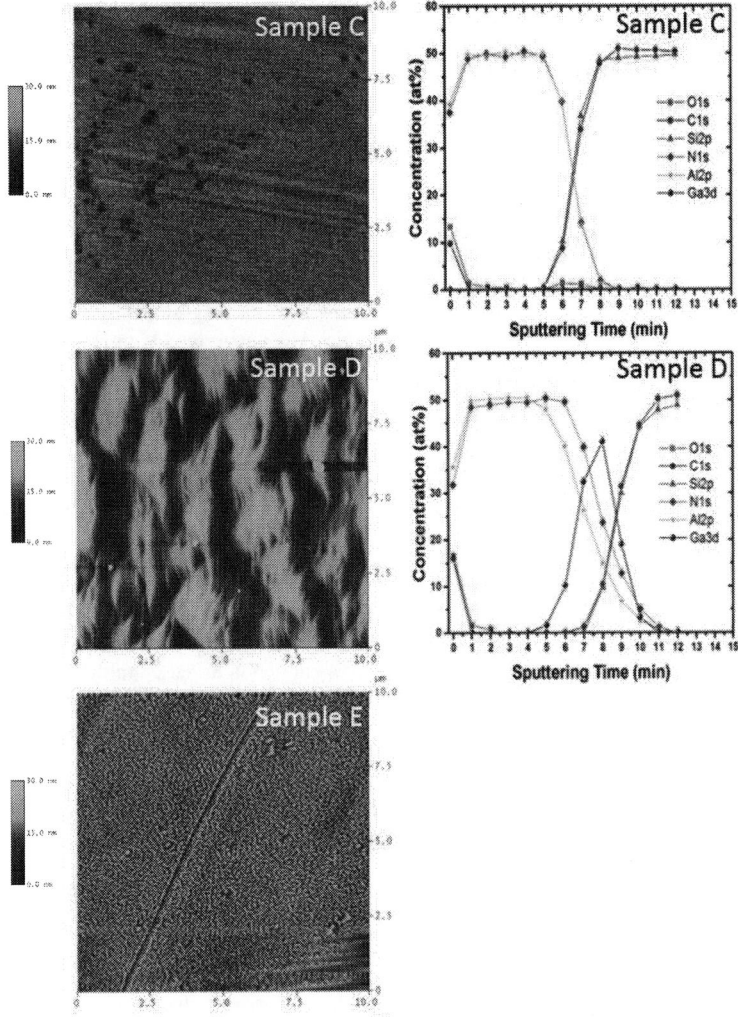

Figure 3. Atomic force microscopy (left) and XPS signatures (right) for in situ substrate preparation (C – plasma off), (D, E – plasma on).

Figure 2 compares SIMS data for samples A and B. The x-axis represents the depth of the ion beam during the measurement, where the origin is the topmost surface of the AlN film and the AlN-SiC interface is approximately 30 nm into each structure. Both samples show evidence of Al diffusion into the SiC as evidenced by a spike in the measured Al signal at the hetero-interface. However, the spike in Al for the film growth by the direct approach appears to be ~3.5 nm deeper than that of the MEE approach, which generally agrees with what is observed XPS. Similarly, both samples show evidence of Si migration into the AlN epilayer with the MEE sample having ~ 1 order of magnitude lower Si concentration over that of the film grown directly. The C profiles for both samples are very similar, with a soft transition at the hetero-interface and a large C concentration observed in the AlN film. Determining whether the source of this C is from diffusion at the hetero-interface or associated with impurities incorporating during growth required more experimentation. However, these results indicate significant intermixing occurs the hetero-interface for both samples independent of growth method.

Figure 3 demonstrates the impact of an active N plasma source during the Ga predeposition process despite closing the source shutter on the morphology and interface composition of the AlN films. Sample E was prepared with the N plasma on and was grown at a substrate temperature of 600 °C. The film exhibits a rough surface morphology despite the use of MEE approach, consistent with the observations of Okumura et al. [5]. Sample D, grown under similar conditions at a higher substrate temperature of 740 °C exhibits a 2D morphology characterized by micro-steps and possible step-bunching. Sample C was also grown at 600 °C similar to sample D, but with the N plasma off during Ga deposition, and exhibits an atomically smooth surface morphology consistent with that of the substrate. These results indicate that the presence of running the N plasma source with the shutter closed during the Ga pre-deposition process results in subsequent 3D nucleation of AlN by reducing the surface mobility of adatoms. Increasing the substrate temperature suppressed this effect resulting in a step-flow growth.

Examination of the XPS measurements of samples C and D provide an indication of the source of this effect (Figure 3 right). XPS study of Sample D shows a clear shift in the drop off of the N1s signal toward further depth with respect to that of the Al2p signal that coincides with a peak in the Ga3d signal. This indicated the presence of GaN at the hetero-interface of this film as well as the possibly metallic Ga based upon the amplitude of Ga 3d signal. Based upon on the growth conditions, this GaN must nucleate during the Ga-pre-deposition process through active N that gets around the plasma source shutter. In contrast, XPS studies of sample C show a sharp and coincident drop-off of the N1s and Al2p signals indicating a sharper AlN/SiC hetero-interface in this sample that is attributed to leaving the N plasma source off during surface preparation. This suggests that the presence of GaN islands prevents the nucleation of AlN in a 2D growth mode at lower substrate temperatures. The possibility that partial nitridation of the SiC is responsible was excluded by realizing smooth AlN films by MEE growth on 4H-SiC without Ga-predeposition (data not shown). Lastly, it is important to note that sample C shows a small but detectable O1s peak at the hetero-interface that is absent in sample D. This may indicate that the lack of active N may also reduce the effectiveness of the Ga pre-deposition process for removing oxides from the substrate surface prior to epitaxial growth.

Conclusion

Migration enhanced epitaxial growth is found to promote 2D nucleation and smoother surface morphology for AlN deposited on 4H-SiC over direct growth. Significant intermixing at the interface and the presence of Si migration from SiC are observed for both growth approaches, although slightly reduced by MEE approach. Active nitrogen exposure during in situ Ga pre-deposition is found to hinder the 2D nucleation of AlN that is attributed to the formation of GaN islands at the hetero-interface.

References

1. Torvik, J. T., Leksono, M., Pankove, J. I., & Van Zeghbroeck, B. (1999). A GaN/4H-SiC heterojunction bipolar transistor with operation up to 300 C. *MRS Internet Journal of Nitride Semiconductor Research, 4*(01), e3.
2. H. Miyake, K. Amari, T. Kimoto, J. Suda, Jap. J. of Appl Phys 52 (2013) 124102
3. A. V. Sampath, R. W. Enck, Q. Zhou, D.C. McIntosh, H. Shen, J.C. Campbell, and M. Wraback, Appl. Phys.Lett., 101, 093506 (2012).
4. Rodak, L. E., Sampath, A. V., Gallinat, C. S., Chen, Y., Zhou, Q., Campbell, J. C., ... & Wraback, M. (2013). Solar-blind AlxGa1− xN/AlN/SiC photodiodes with a polarization-induced electron filter. *Applied Physics Letters, 103*(7), 071110.
5. H. Okumura, T. Kimoto and J. Suda , Applied Physics Express 4 025502 (2011).
6. Hoke, W. E., et al. "Rapid silicon outdiffusion from SiC substrates during molecular-beam epitaxial growth of AlGaN/ GaN/ AlN transistor structures." *Journal of applied physics* 98.8 (2005): 084510.

Chapter 2

Nitride Devices

GaN MIS-HEMT with Low Dynamic ON-resistance Using SiON Passivation

S. C. Liu[a], C. K. Huang[b], and E. Y. Chang[a,b]

[a] Department of Materials Science and Engineering, National Chiao-Tung University, Hsinchu, 300 Taiwan.
[b] Department of Electronics Engineering, National Chiao-Tung University, Hsinchu, 300 Taiwan.

An effective SiON passivation with high density of positive fixed charges for GaN MIS-HEMTs is demonstrated. The positive fixed charges at the SiON/AlGaN interface effectively reduce the surface potential and expand the quantum well below the Fermi level, thus improving the device performance. The GaN MIS-HEMT with SiON demonstrated a high maximum drain-source current density ($I_{DS,max}$) of >1 A/mm, a breakdown voltage of 750 V at a drain leakage current of 1 µA/mm, and a well transfer characteristics. The dynamic ON-resistance only increased slightly under a high quiescent bias of 100 V.

Introduction

Gallium nitride-based high-electron-mobility transistors (GaN HEMTs) have demonstrated outstanding performance for high-power and high-frequency applications for defense and communication systems. However, there are many undesirable effects such as current collapse and the increase of dynamic ON-resistance due to the surface states and the high polarization nature of the GaN-based material [1], [2]. SiN passivation proved to reduce the density of surface states can effectively mitigate current collapse and the increase of dynamic ON-resistance [3]. However, the large amount of interface states at SiN/GaN interface still lead to the current collapse when the GaN HEMT operate under high-electric field. In this work, we demonstrated an effective SiON passivation with high density of positive fixed charges for GaN MIS-HEMTs which can. The high density of positive fixed charges at the SiON/GaN interface can compensate the negative charges and reduce the negative GaN surface potential, thus expanding the 2DEG quantum well below the Fermi level and stabilizing the 2DEG carrier density when the device operates under high-electric field.

Experiments

The AlGaN/GaN HEMT heterostructure was grown by metal-organic chemical vapor deposition on silicon substrate. The epitaxial structure consisted of 1-nm GaN cap layer, 25-nm $Al_{0.2}Ga_{0.8}N$ barrier layer, 1.3-µm i-GaN layer and a buffer layer consisted of GaN/AlGaN/AlN with total thickness of 4-µm. The wafer was divided into two samples after mesa and ohmic contact process. The nitrogen passivation technique was adopted to recover and clean the GaN surface prior to passivation [4]. The passivation layer were prepared differently for each sample: sample A with 12-nm SiON, sample B with 12-nm SiN. The gate-to-drain spacing L_{GD}, gate-to-source spacing L_{GS}, and gate length L_G were 10-µm, 3-µm, and 2-µm, respectively. The schematic cross section of the AlGaN/GaN HEMT with passivation layer and gate insulator is shown in Fig. 1.

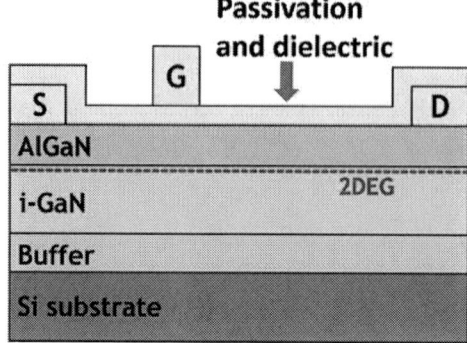

Figure 1: Schematic cross section of the GaN MIS-HEMT.

Results and Discussion

For the transfer characteristics shown in figure 2(b), the negatively shift of V_{th} indicates the existence of positive fixed charge between passivation layer and AlGaN interface [4]. The density of positive fixed charges were calculated to be ~2.7 × 10^{13} and ~1.5 × 10^{13} e/cm^{-2} for SiON and SiN, respectively. The basic DC *I–V* characteristics are shown in Fig. 2. For the sample with SiON passivation, a higher $I_{DS,max}$ of >1 A/mm, and a lower subthreshold slope of 68 mV/dec were obtained. In contrast, the sample with SiN passivation exhibits a $I_{DS,max}$ of ~896 mA/mm, and a subthreshold slope of 73 mV/dec. It indicates that the positive fixed charges at the interface between passivation layer and AlGaN interface reduce the negative GaN surface potential and expand the quantum well below the Fermi level, resulting in the increase of the 2DEG carrier density. The dynamic ON-resistance has been commonly used to examine the trapping effects attributed to the surface and interface states in the GaN structure [5]. As shown in Fig. 3, the dynamic ON-resistance were extracted by various OFF-state drain quiescent voltage of 0 to 100 V and ON-state with $V_{GS} = 0$ V, $V_{DS} = 1$ V. The ON-state pulse width was 500 µs with a duty cycle of 10%. For the V_{DSQ} stress at 100 V, the dynamic ON-resistance increases slightly to 1.03 times for the sample with SiON passivation. In contrast, the dynamic ON-resistance increases 1.17 times for the sample with SiN passivation. The results reveal that SiON passivation with high density of positive fixed charge for GaN MIS-HEMT is preferable for power device applications.

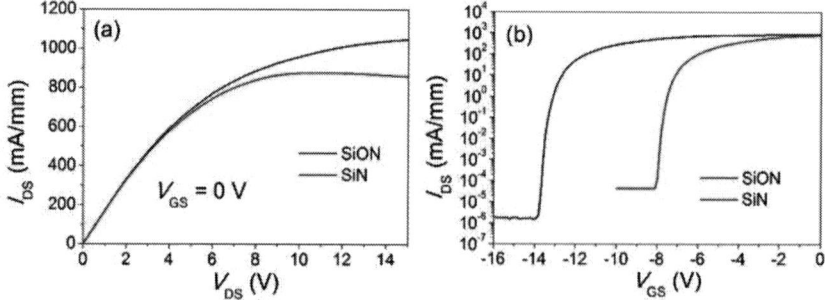

Figure 2: (a) I_{DS}–V_{DS} characteristics and (b) transfer characteristics for GaN MIS-HEMTs with different passivation and gate insulator layers.

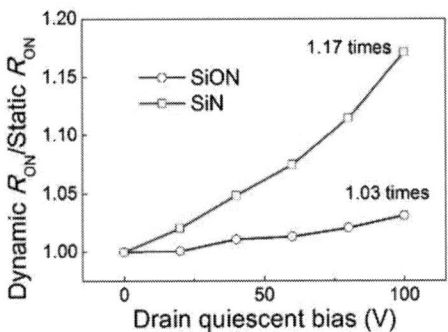

Figure 3: Switching performance extracted from various OFF-state quiescent bias.

Summary

SiON passivation with high density of positive fixed charges is proved to improve the performance of GaN MIS-HEMTs. With SiON passivation, the GaN MIS-HEMT exhibits significant improvements in I–V characteristics and dynamic ON-resistance compared to the conventional SiN passivated device. Overall, SiON with high density of positive fixed charges is a promising passivation for GaN power devices.

Acknowledgement

This work was sponsored by the TSMC, NCTU-UCB I-RiCE program, and Ministry of Science and Technology, Taiwan, under Grant No. MOST 105-2911-I-009-301 and National Chung-Shan Institute of Science & Technology, Taiwan, under Grant No. NCSIST-102-V211(105).

References

1. W. Saito, Y. Takada, M. Kuraguchi, K. Tsuda, I. Omura, T. Ogura and H. Ohashi, IEEE Transactions on Electron Devices, 50 (12), 2528-2531 (2003).
2. R. Vetury, N. Q. Zhang, S. Keller and U. K. Mishra, Electron Devices, IEEE Transactions on 48 (3), 560-566 (2001).
3. Z. Zhuo, Y. Sannomiya, Y. Kanetani, T. Yamada, H. Ohmi, H. Kakiuchi and K. Yasutake, Nanoscale research letters 8 (1), 1-6 (2013).
4. G. Dutta, S. Turuvekere, N. Karumuri, N. DasGupta, and A. DasGupta, IEEE Electron Device Letters, 60 (11), 1085-1087 (2014).
5. D. Jin and J. del Alamo, IEEE Transactions on Electron Devices, 60 (10), 3190-3196 (2013).

22

Improved GaN based Hydrogen Sensors

Kwang Hyeon Baik[a], Jimin Kim[b], Soohwan Jang[b]

[a]Department of Materials Science and Engineering, Hongik, University, Jochiwon, 30016, Republic of Korea
[b]Department of Chemical Engineering, Dankook University, Yongin, 16890, Republic of Korea

To improve the conventional GaN based hydrogen sensor devices, catalytically active Pt nanonetworks were applied to active gate area of AlGaN/GaN HEMT, and surface roughening of active area in nonpoalar a-plane and semipolar GaN diode by using Photo-electrochemical etching was employed. When the active gate region of an AlGaN/GaN sensor is functionalized with platinum nanostructures that contain a larger surface area offering more active sites for hydrogen molecules to be adsorbed, the drain current response was dramatically improved. Also, the extended rough surface of the diodes showed improved hydrogen detection sensitivity due to the presence of more available adsorption sites, resulting in effective variations of the Schottky barrier height.

Introduction

There are great interests in hydrogen as one of the most promising alternative energy source and carrier. Hydrogen is emission-free fuel when reacted with oxygen without producing any harmful byproducts such as carbon dioxide and nitrogen oxide. Recently, great efforts have been devoted to produce hydrogen in efficient way such as photoelectrochemical water splitting, photocatalytic water splitting, and photobioligical water splitting. Hydrogen gas is colorless, odorless, extremely reactive with oxygen, and has low ignition energy [1]. Since it has negative Joule-Thomson coefficient, hydrogen gas leaking from pressurized container raises its temperature. This may induce spontaneous flammable ignition. Therefore, prompt and reliable hydrogen gas detection is very important in various hydrogen related industrial processes and equipments for the safety.

GaN based material is very suitable compound semiconductor to build up hydrogen sensor system. Wide bandgap of GaN enable the sensors to be operated at high temperature and harsh radiative environment, and mechanical and chemical robustness of GaN guarantees the reliability and durability of the devices [2, 3]. Many types of GaN based hydrogen sensing devices including Schottky diode, metal oxide semiconductor (MOS) diode, GaN nanowire and AlGaN/GaN high electron mobility transistors (HEMTs) which employ catalytically active platinum or palladium film layers on the gate region have been developed for fast and sensitive detection of hydrogen [4-9]. Among them, AlGaN/GaN HEMT structure with 2 dimensional electron gas channel (2-DEG) induced by piezoelectric and spontaneous polarization between the AlGaN and GaN layers shows highly sensitive current changes to surface charges created by catalytic reaction of hydrogen molecules on active layer [4, 10].

In this research, AlGaN/GaN HEMT hydrogen sensors with Pt nanostructure and GaN diode sensors with etched surface were introduced as an approach to improve the sensitivity of hydrogen sensors. Instead of planar Pt film, the sensitivity of HEMT with catalytic Pt nanostructure on the gate region was dramatically improved. Also, a-plane (11-20) (a-GaN) and semipolar (11-22) GaN (s-GaN) based diode with etched large surface area showed excellent hydrogen response by offering more available adsorption sites to hydrogen molecules.

Experiments

Pt nanonetwork with 2-3 nm diameter which is random connection of nanorods in nanostructure was synthesized by a simple solution phase method at room temperature [11]. Synthesized nanonetworks were applied to the semiconductor surface by spin coating. Density and distribution of networks on the surface were controlled by the number of spin coating cycles. Pt nanonetwork was deposited on the gate region of AlGaN/GaN HEMT selectively using conventional and facile lift-off technique. Once nanonetworks were attached on the solid surface, they would not be detached during the following semiconductor fabrication processes such as cleaning and lift-off.

For etched a-GaN and s-GaN diode fabrication, the Ohmic metal stack was deposited by e-beam evaporation, patterned by lift-off, and annealed. A 200 nm thick Si_3N_4 passivation layer was formed for diode isolation using plasma enhanced chemical vapor deposition. The windows for the active sensing area opening were achieved through buffered oxide etchant etching. The samples were immersed in 1 M potassium hydroxide (KOH) solution. Photo-electrochemical (PEC) etchings were performed at a stirring rate of 300 rpm at 80°C with ultraviolet illuminations. After the PEC etching of Schottky contact region, a 10 nm Pt film was evaporated following Ti/Au contact pads for probing and wire bonding.

Current–voltage characteristics of the devices exposed to hydrogen balanced with dry air were measured at room temperature in a gas test chamber using an Agilent 4155C semiconductor parameter analyzer.

Results and Discussion

To improve the sensitivity of AlGaN/GaN HEMT based hydrogen sensors, a nanostructured platinum layer with a large surface to volume ratio can be employed [12, 13] Pt nanoparticles with a large surface area can improve catalytic reactivity, as well as overcome high cost and limited supply concerns in commercial usage. Moreover, when the active gate region of an AlGaN/GaN sensor is functionalized with platinum nanostructures that contain a larger surface area offering more active sites for hydrogen molecules to be adsorbed, the drain current response can be dramatically improved.

Figure 1(a) shows the drain current-voltage characteristics of the HEMT with platinum nanonetworks in both air and 500 ppm H_2 ambient conditions. It is important to note that the hydrogen induced drain current increases of the nanonetwork sensor were significantly larger than that of conventional platinum film HEMT. After exposure to 500 ppm hydrogen, the drain current increase at V_{DS} of 6 V and V_{GS} of 0V was 4.5 mA for the nanonetwork HEMT. When hydrogen molecules are introduced to the HEMT, they reach the platinum nanonetwork surface and are decomposed into hydrogen ions. Then, the dissociated hydrogen ions diffuse into the AlGaN interface to form effective positive charges, thereby improving the 2DEG channel and increasing the drain current. Platinum nanonetworks grown by simple solution synthesis can have a huge surface area of $53m^2/g$ [14]. Subsequently, a larger number of hydrogen molecules can be absorbed which react with the abundant catalytic sites on the platinum thin film. Relative current changes of platinum nanonetwork and film sensors according to gate bias voltages at V_{DS} of 6 V for 500 ppm hydrogen gas exposure are shown in figure 1 (b). The maximum percentage of current change for the platinum nanonetwork sensor is 3.3×10^6 % at V_{GS} of -3.3 V, while just 2.5×10^2 % at V_{GS} of -2.9 V for the Pt film. The AlGaN/GaNHEMT with platinum nanonetworks has a notably improved current response to hydrogen due to the larger active surface area of the nanostruture.

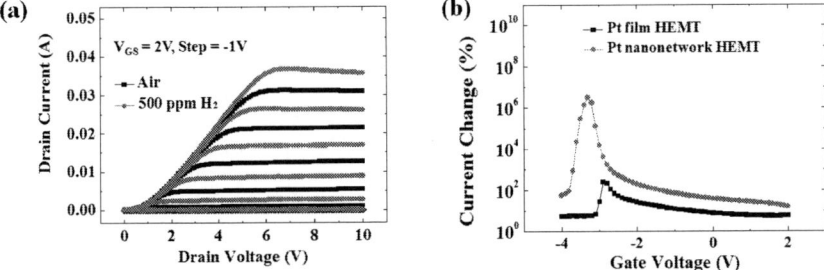

Figure 1. (a) Drain current voltage characteristics of Pt nanonetwork gated HEMT. (b) Relative current change of Pt nanonetwork gated HEMT measured at drain voltage of 6 V under air and 500 ppm H_2 ambient.

The hydrogen sensitivity of GaN-based hydrogen sensors can be improved not only by functionalizing the active gate area containing nanostructures, but also by incorporating surface etching on the active contact area using wet chemical solutions. Figure 2 presents SEM images of the etched surface morphology of the Schottky contact area on nonpolar a-GaN (a), (c) and semipolar s-GaN (b), (d) after 7 min of PEC wet etching using the KOH solution. Striated trigonal prisms with submicron widths are observed on the top surface of the a-GaN film. In addition, inclined trigonal unit cells are exposed on the s-GaN films. In both cases, the specific crystallographic planes of {100} and (0001) were similarly exposed, indicating the planes were chemically stable due to atomic bond configurations and a smaller density of the atoms. By using a rough surface, a larger number of absorption sites could be created which in turn helped improve overall hydrogen detection sensitivity. Thus, as more hydrogen atoms are induced on the surface, changes in the Schottky barrier height can be achieved.

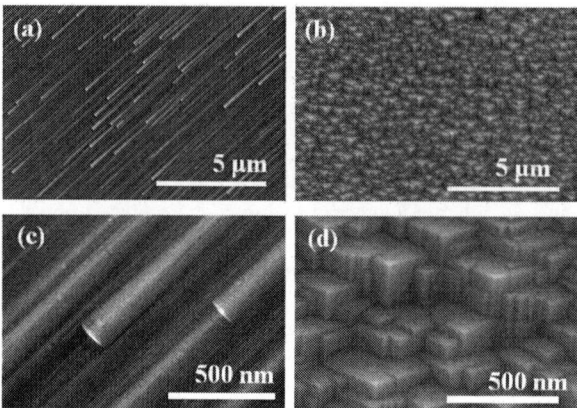

Figure 2. SEM images of etched surface for (a), (c) nonpolar a-GaN and (b), (d) semipolar s-GaN.

Figure 3. shows the current-voltage characteristics for the Pt Schottky diodes on nonpolar a-GaN and semipolar s-GaN films with and without wet etching before and after exposure to 4% H_2 in N_2 at room temperature. Both Pt Schottky diodes showed remarkable current changes when exposed to hydrogen in the sweeping bias range. Clearly, the surface-etched Pt Schottky diodes exhibited a better response to H_2 than the non-etched ones. The benefit of this approach is that the surface polarity and atomic configurations are advantageous when responding to H_2 molecules, possibly due to a much higher affinity of hydrogen to nitrogen in the GaN surface. Specifically, a large density of exposed neighboring nitrogen atoms with dangling bonds exists both on the nonpolar GaN surface and beneath the semipolar GaN surface. This, in turn, can effectively enhance variations in the Schottky barrier height.

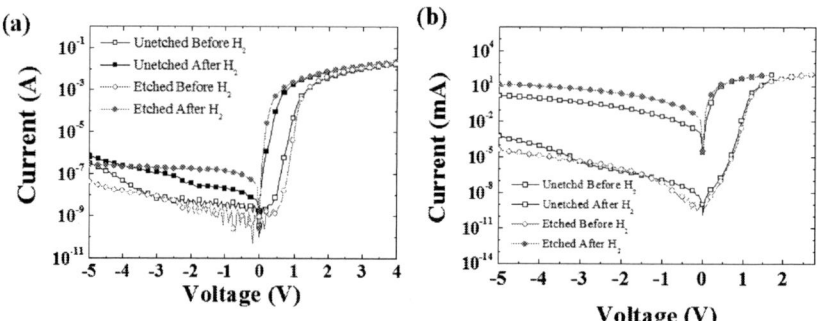

Figire 3. The current-voltage characteristics for the Pt Schottky diodes on (a) nonpolar a-GaN and (b) semipolar s-GaN films with wet etching before and after exposure to 4% H_2.

Figure 4. shows the relative current change to hydrogen exposure as a percentage value for the Pt Schottky diodes with and without wet etching. The etched Schottky diodes showed a larger current change over the sweeping bias voltages than the non-etched one, especially one or two orders of magnitude higher in the reverse bias range. It also confirms that the etched Schottky diodes exhibited a larger current response when exposed to an H_2-containing ambient brought on by a decrease in effective barrier lowering. The maximum hydrogen sensitivity of the etched diode is 6-7 times higher than that of the non-etched one. The hydrogen gas sensors showed stable and reproducible current changes, as well as the ability to cycle this current in response to repeated introductions of hydrogen into the ambient. These results suggest that hydrogen detection sensitivity can be improved by using surface etching with KOH solutions for Pt Schottky diodes fabricated on nonpolar a-GaN and semipolar s-GaN films.

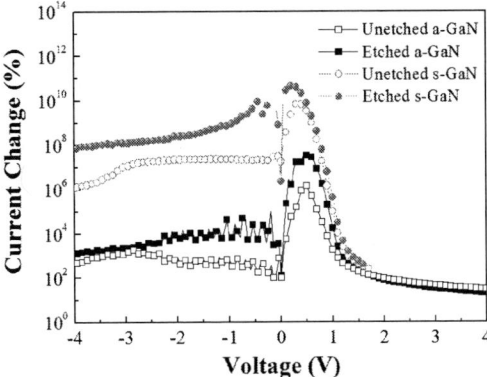

Figure 4. The relative current change to hydrogen exposure for the Pt Schottky diodes with and without wet etching for a-GaN and s-GaN Schottky diodes.

Conclusions

As an effective approach to improve the conventional GaN based hydrogen sensor devices, Pt nanonetworks were applied to active gate area of AlGaN/GaN HEMT. When the active gate region of an AlGaN/GaN sensor is functionalized with platinum nanostructures that contain a larger surface area offering more active sites for hydrogen molecules to be adsorbed, the drain current response was dramatically improved. Also, surface roughening of active area in nonpoalar a-plane and semipolar GaN diode by using PEC etching was employed. The extended rough surface of the diodes showed improved hydrogen detection sensitivity due to the presence of more available adsorption sites, resulting in effective variations of the Schottky barrier height.

Acknowledgments

This research was supported by Basic Science Research Program through the National Research Foundation of Korea (NRF) funded by the Ministry of Education(2014R1A1A4A01008877, 2015R1D1A1A01058663), and Nano · Material Technology Development Program through the National Research Foundation of Korea (NRF) funded by the Ministry of Science, ICT and Future Planning(2015M3A7B7045185, 2009-0082580).

References

1. G.R. Astbury and S.J. Hawksworth, *Int. J. Hydrogen Energy*, **32**, 2178 (2007).
2. L. Voss, B.P. Gila, S.J. Pearton, H. Wang, and F. Ren, *J. Vac. Sci. Technol. B* **23**, 2373 (2005).
3. H. Wang, T.J. Anderson, F. Ren, C. Li, Z. Low, J. Lin, B.P. Gila, S.J. Pearton, A. Osinsky, and A. Dabiran, *Appl. Phys. Lett.* **89**, 242111 (2006).
4. S. J. Pearton, B. S. Kang, S. Kim, F. Ren, B. P. Gila, C. R. Abernathy, J. Lin and S. N. G. Chu, *J. Phys. Condens. Matter* **16**, R961-R994 (2004).
5. C. Chang, G.C. Chi, W. Wang, L. Chen, K. H. Chen, F. Ren and S. J. Pearton, *J. Electron. Mater.* **35**, 738 (2006).
6. W. Lim, J. S. Wright, B. P. Gila, J. L. Johnson, A. Ural, T. Anderson, F. Ren, and S. J. Pearton, *Appl. Phys. Lett.* **93**, 072110 (2008).
7. H.-T. Wang, T. J. Anderson, B. S. Kang, F. Ren, C. Li, Z.-N. Low, J. Lin, B. P. Gila, S. J. Pearton, A. Osinsky and A. Dabiran, *Appl. Phys. Lett.* **90**, 252109 (2007) .
8. H.T. Wang, B.S. Kang, F. Ren, R.C. Fitch, J. Gillespie, N. Moser, G. Jessen, R. Dettmer, B.P. Gila, C.R. Abernathy and S.J. Pearton, *Appl. Phys. Lett.* **87**, 172105 (2005) .
9. B.S. Kang, R. Mehandru, S. Kim, F. Ren, R. Fitch, J. Gillespie, N. Moser, G. Jessen, T. Jenkins, R. Dettmer, D. Via, A. Crespo, B.P. Gila, C. R. Abernathy and S. J. Pearton, *Appl. Phys. Lett.* **84**, 4635-4637 (2004).
10. H. Kim, and S. Jang, *Curr. Appl. Phys.* **13**, 1746 (2013).
11. Y. Song, R. M. Garcia, R. M. Dorin, H. Wang, Y. Qiu, E. N. Coker, W. A. Steen, J. E. Miller, and J. A. Shelnutt, *Nano Lett.* **7**, 3650 (2007).
12. A. Rouxoux, J. Schulz, and H. Patin, *Chem. Rev.* 102, 3757 (2002)
13. Z. Peng, and H. Yang, *Nano Today*, **4**, 143 (2009)
14. Y. Jung, J. Ahn, K. H. Baik, D. Kim, S. J. Pearton, F. Ren and J. Kim, *J. Electrochem. Soc.* **159,** H117 (2012)

Chapter 3

Nitride Optoelectronics

30

On the origin of the 4.7 eV absorption and 2.8 eV emission bands in bulk AlN substrates

D. Alden[a,b], Z. Bryan[a], B. E. Gaddy[a], I. Bryan[a], G. Callsen[b], A. Koukitu[c], Y. Kumagai[c], A. Hoffmann[b], D. L. Irving[a], Z. Sitar[a], and R. Collazo[a]

[a] Department of Materials Science and Engineering, North Carolina State University, Raleigh, NC 27695-7919, USA
[b] Technical University Berlin, Solid State Physics Institute, Hardenbergstr. 36, 10623 Berlin, Germany
[c] Department of Applied Chemistry, Tokyo University of Agriculture and Technology, Koganei, Tokyo 184-8588, Japan

One of the main limitations based on point defects in AlN is related to the UV absorption band present at 265 nm. This relatively broad absorption band limits the use of bulk substrates within the deep UV range and several complicated fabrication procedures have been devised to overcome this limitation. The origin for this absorption band along with corresponding photoluminescence signatures will be reviewed. C_N^- and V_N^+ were identified as the point defects directly associated with the main absorption at 4.7 eV and the emission at 2.8 eV. The main absorption is due to a transition between the C_N^- state and conduction band, while the emission arises from a DAP transition between the donor V_N^+ and acceptor C_N^- through the 4.7 eV excitation channel. New observations based on photoluminescence excitation are presented to further support the assignments previously made.

Introduction

AlGaN alloys are the building blocks for deep UV optoelectronics and high-power devices (1-3). It has been demonstrated that the highest crystalline quality AlGaN films with high Al content are obtained on AlN single crystal substrates (4). Pseudomorphic AlGaN films with Al content higher than 50%, and dislocation densities lower than 10^4 cm^{-2} have been achieved on these substrates, sustaining compressive stresses with thicknesses exceeding 3 μm. Such results demonstrate the advantages of using AlN substrates for this technology, and at such some have been realized on several deep-UV optoelectronics applications (5). Inoue et al. have demonstrated UV LEDs emitting at 265 nm with output powers exceeding 80 mW using hydride vapor phase epitaxy (HVPE) grown AlN on bulk AlN substrates (6). UV LEDs grown on these native substrates have higher reliabilities and higher output powers. In addition, optically pumped lasers emitting at wavelengths between 230 nm and 280 nm that display cavity modes and single polarized-state emission with low lasing thresholds have been developed (7, 8). Nevertheless, there are several limitations related to the performance of these devices.

These limitations are classified in two main categories: (1) identification and control of point defects, and (2) efficient doping.

One of the main limitations based on point defects is the one related to the UV absorption band present at 265 nm in AlN (9-11). This relatively broad absorption band limits the use of the substrate within the deep UV range and several complicated fabrication procedures were devised to overcome this limitation. Several publications have recently addressed this topic, taking advantage of new developments on methods for predicting defect states using density functional theory (DFT). In this article, the origin for this absorption band along with corresponding photoluminescence (PL) signatures will be reviewed. Furthermore, new observations based on photoluminescence excitation (PLE) will be presented and discussed to further support the assignments made in previous publications.

A strong optical absorption band below bandgap in the blue/UV range has been present in single crystalline AlN for over 15 years (9, 10, 12-14). Not only do these absorption bands limit the full potential of using AlN substrates for optoelectronic applications but they also indicate the presence of point defects that would limit the full potential of high thermal conductivity for electronic device applications. The absorption band in which we are most interested in is centered near 4.7 eV (~265 nm) and can have absorption coefficients well over 1000 cm^{-1}. A first attempt at understanding the origin of this absorption band was made by Slack et al. where it was determined that there was a strong correlation with oxygen concentration, where the minimum oxygen concentration in their material was 5×10^{19} cm^{-3} (15). The samples in these studies have oxygen concentrations well below this amount, which contrasts to the initial identification.

Experimental

The AlN single crystal wafers used in this study were grown by physical vapor transport (PVT) using tantalum carbide crucibles in an inductively heated reactor (16-18). Seeded growth was realized on N-polar c-plane ($000\bar{1}$) seeds. Wafers obtained from grown boules were around 500 μm thick with high crystalline quality. As carbon was recognized as an impurity of interest, it was necessary to develop alternative methods to achieve its control. In this particular set of experiments, in order to achieve this control, thick homoepitaxial films were grown by atmospheric pressure HVPE on the AlN single crystal wafers (19, 20). The crystalline quality of these films, as assessed by x-ray diffraction, was similar to that of the original bulk AlN wafers . After HVPE growth, the final film thicknesses exceeded 180 μm, such that after grinding off the substrate and optically polishing both sides of the HVPE layers yielded stand-alone HVPE wafers that were approximately 100 μm thick.

Impurity concentrations within the single crystal wafers were determined by secondary ion mass spectrometry (SIMS). Photoluminescence measurements were realized at room temperature by using a 193 nm ArF$_2$ laser as source and a 0.75 m Princeton Instrument spectrograph/monochromator with a UV sensitive CCD detector. Special attention was taken to maintain a similar optical path and illumination area in order to allow for the possibility of intensity comparison between the different samples. Photoluminescence excitation (PLE) mapping was realized using as source a Xenon arc lamp. The excitation wavelength was selected using a grating monochromator and was then guided onto the sample. The photoluminescence spectra were measured at room temperature utilizing a grating monochromator and a liquid nitrogen cooled CCD detector.

Density functional theory (DFT) provides a route of solving the many body electronic structure problems in terms of the electron density. Although DFT is formally exact, the form of the exchange correlation functional is unknown and is approximated. Once the exchange correlation functional is approximated, implementation of DFT enables the prediction and understanding of the atomic and electronic structure of materials. More specifically, in this work, DFT was used to calculate defect formation energies as well as thermodynamic and optical transition levels. The DFT was performed using a supercell consisting of 96 atoms computed with a 2 x 2 x 2 Monkhorst-Pack k-points grid with an energy cutoff of 500 eV and a plane wave basis set, as implemented in the Vienna *ab initio* simulation package. By using traditional exchange-correlation functionals, the bandgap is typically under predicted; therefore, as a novel approach in these computational methods, the Heyd-Scuseria-Ernzerhof screened hybrid exchange functional was used, which mixes Hartree-Fock (HF) exchange in the short range with the Perdew-Burke-Ernzerhof exchange correlation functional. The amount of short-range HF exchange was set to 0.32 to match best the experimental bandgap and lattice constants. With these calculations, the most probable defects to form can be predicted. In combination with experimental results, DFT is a powerful tool for building models and hypothesis useful for point defect studies.

Results and Discussion

C_N identification

The initial approach to the problem was to use SIMS measurements to identify the three most common impurities: silicon, oxygen, and carbon. The typical amount of carbon present in the substrates is 2×10^{19} cm^{-3}, while silicon and oxygen levels are below 2×10^{18} cm^{-3}. This strongly suggested the significant role carbon might play in the strong absorption band centered on 4.7 eV, thus providing a starting point for predictive DFT calculations. Multiple carbon-based defects were considered and calculations performed to determine the absorption and emission transitions between the defects. Of all the defects studied, only substitutional carbon on a nitrogen site resulted in a predicted absorption transition around 4.7 eV (12).

The formation energy for a defect on a given site in a charge state q is the difference between the energy of the defective system to a reference bulk system. For the case of carbon substituting on a nitrogen site, the formation energy is determined by the following expression,

$$E^f\left(C_N^q\right) = E_{tot}\left(C_N^q\right) - E_{tot}\left(bulk\right) - \mu_C + \mu_N + q\left(E_F + VBM + \Delta V\right) \qquad [1].$$

The total energy terms are calculated using DFT. The chemical potential of nitrogen is limited on the upper end by the formation free energy of gas phase nitrogen and on the lower end by the enthalpy of formation of AlN, thus $\mu_{Al} + \mu_N = \Delta H_{AlN}^f$. The chemical potential of carbon is based on the energy of bulk diamond and the ΔV is a term used to align the reference potential in the charged unit cell to the uncharged bulk. The formation energy as a function of Fermi energy for substitutional carbon point defects represented at the nitrogen rich (a) and aluminum rich (b) limits of the chemical potential is given in Figure 1 (21).

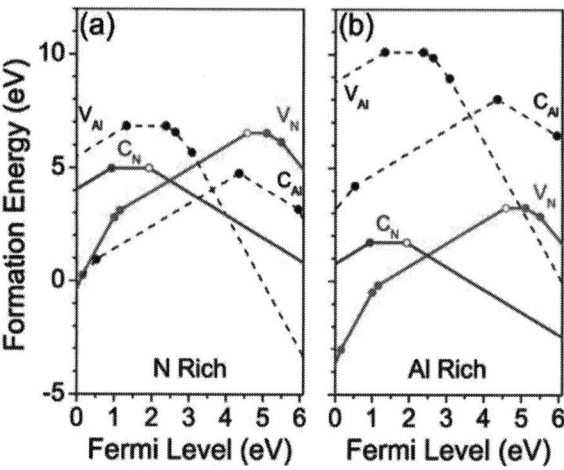

Figure 1. Formation energy for carbon and vacancy defects in AlN for nitrogen rich (a) and aluminum rich (b) growth conditions (21).

Al and N vacancy formation energies are also included. A given point defect shows a thermodynamic preference for the charge state that minimizes the formation energy. Therefore only the minimum formation energy for each charge state is included where the transition point between two charge states corresponds to equal formation energies for both defects. Figure 1 shows that the formation energy of substitutional carbon on an aluminum site is only favorable near the nitrogen rich condition extreme; otherwise carbon on a nitrogen site has the more favorable energy of formation (21). For Fermi energies higher than 2 eV above VBM, carbon on the nitrogen site will likely be in the -1 charge state. In this state, it acts as an acceptor with a thermodynamic transition level of 4.2 eV. The optical transitions corresponding to the transitions between C_N^- and $C_N^0 + e^-$ are shown in Figure 2 (12).

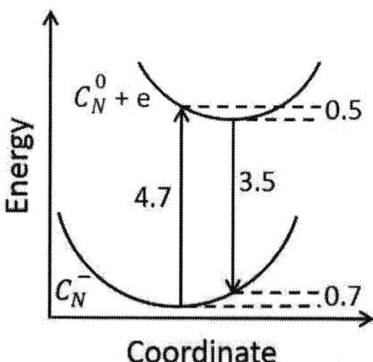

Figure 2. Configuration coordinate diagram for substitutional carbon on the nitrogen site in AlN. Energies are in eV.

The model predicts an absorption around 4.7 eV arising from the C_N^- state to the conduction band. After relaxation, an emission back to the C_N^- state will have a minimum energy of 3.5 eV. The vertical transitions are calculated using the Franck-Condon approximation, which assumes that the optical transition occurs faster than the subsequent lattice relaxation. The sum of the Franck-Condon shifts yields a maximum Stokes shift of 1.2 eV.

In order to experimentally verify this model, thick HVPE AlN films with different carbon concentrations ranging from $< 2 \times 10^{17}$ cm^{-3} to 1×10^{19} cm^{-3} and a PVT AlN crystal with carbon concentration of 2×10^{19} cm^{-3} were grown. As mentioned in the previous section, absorption measurements were realized by grounding off the substrate in addition to achieving an optical polish surface on both sides. The uncorrected (to reflection) absorption coefficient as a function of incident energy for the samples is shown in Figure 3 (12). As predicted by the model, there is a clear correlation between the absorption coefficient around 4.7 eV and carbon concentration. The observed absorption energy closely agrees with the DFT model since the absorption process is likely to originate from the vibrational minimum. The absorption coefficient is plotted for each carbon concentration in Figure 3 (b) where a linear relationship between carbon concentration and the absorption band exists. This relationship follows the Fermi Golden Rule where the absorption coefficient is linearly proportional to the number of states available for the transition assuming that the majority of carbon is incorporated in the C_N^- state.

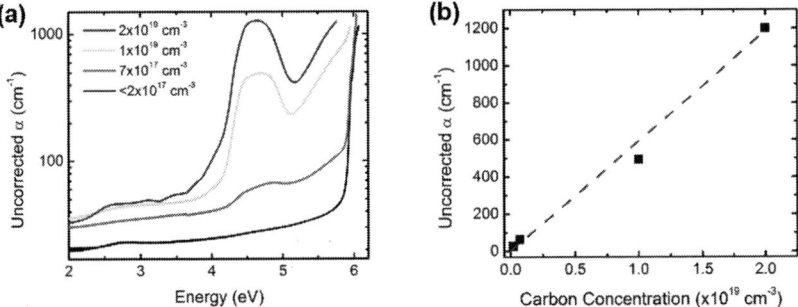

Figure 3. Uncorrected absorption coefficient as a function of incident photon energy for single crystalline AlN with different carbon concentrations (a). Uncorrected absorption coefficient at 4.7 eV as a function of carbon concentration (b).

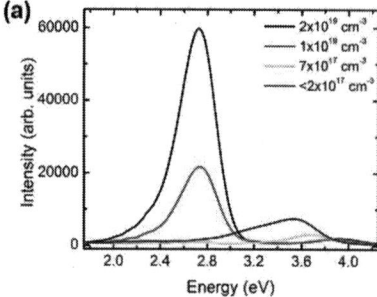 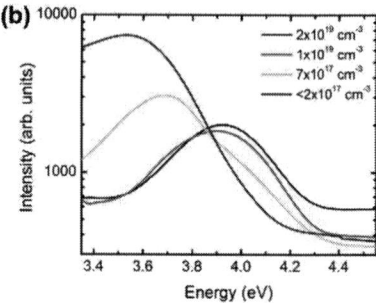

Figure 4. Relative PL spectra at RT for single crystalline AlN with different carbon concentrations (a). The 3.9 eV emission intensity dependence on carbon concentration (b).

Relative photoluminescence measurements taken at room temperature are shown in Figure 4. A closer look of the emission intensity around 3.9 eV is shown in Figure 4 (b). These measurements show that the emission intensity 3.9 eV can be directly correlated with the carbon concentration, where at the lowest carbon concentration this emission is not present. At the lower carbon concentrations, emissions with energies centered between 3.3 to 3.6 eV become dominant, as seen in both doped c-plane and m-plane AlN films. These emissions were related to possible transitions between Si, O, Si-complex, and/or O-complex and another defect state or from the conduction band to one of these states. Although others have seen similar emissions in AlN, no definitive assignment has been made. The emission peaked at 3.9 eV was observed at 0.4 eV higher in energy than predicted. The model predicts the lowest possible energy for the emission, as the emission is possible from any other excited vibrational state through partial lattice relaxation and not necessarily from the minimum energy atomic configuration.

V_N identification

In addition to the 3.9 eV emission, there is an observed dependence of the 2.8 eV emission with carbon concentration as seen in Figure 4 (a), as also reported by Nagashima et al. where it was suggested that carbon was indirectly related (22). The transition cannot be explained with a simple free-to-bound transition involving carbon, as is the case for the 3.9 eV emission. This leaves three other possible models: emission from a native defect state that is incorporated to compensate the carbon, emission from carbon related complexes, and emission related to a DAP transition between a deep donor and the acceptor carbon.

The formation energies as a function of Fermi energy for nitrogen and aluminum vacancies are shown in Figure 1 (21). It is expected that typical growth conditions for either the bulk substrates or the HVPE layers are near Al rich conditions as the equilibrium partial pressure of Al during growth is not far from that at equilibrium with is condensed phase. Near this condition, nitrogen vacancies are most favorable to form. For a Fermi level position between 2 eV and 4.5 eV, the nitrogen vacancy is more likely to be found in the +1 charge state, V_N^+. It is expected that this ionized donor is incorporated in high concentrations to compensate the carbon, however a free-to-bound transition involving only V_N^+ would correspond to an emission near 5 eV. In addition to the defects described in Figure 1, fourteen other carbon-containing complexes were evaluated by

DFT and it was found that no carbon-containing complex had an optical signature around 2.8 eV. Thus based on our DFT model, a DAP transition seems to be the likely origin to the 2.8 eV emission. It should be noted however that oxygen and silicon concentrations are at or below 5×10^{17} cm^{-3} in these samples, therefore other donors must be considered. Related work has shown that when the oxygen level in the crystal is near or above the carbon level, the emission at 2.8 eV is not present suggesting that oxygen is not likely to be involved in the origin of this emission.

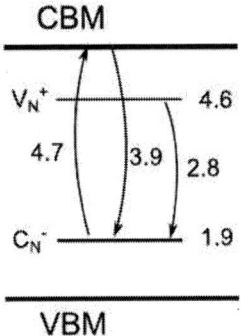

Figure 5. Predicted and observed transitions between the C_N^- state, the V_N^+ state and conduction band minimum (CBM) including Stokes shifts. Energies are in eV.

Based on the formation energies for the defects considered, a vacancy concentration increase is expected with an increase in carbon concentration. If the 2.8 eV emission is related to these two species, the observation of an intensity dependence on the 2.8 eV emission with the carbon concentration is also expected. The energy of the DAP transition is determined by the following equation

$$E_{DAP} = \left(E_g - E_A - E_D \right) + \left[\sqrt[3]{\frac{4\pi}{3}} \frac{e^2}{4\pi\varepsilon\varepsilon_0} \right] N_D^{1/3} \qquad [2]$$

where the bandgap and dielectric constant is assumed to be 6.1 eV and 8.5, respectively, E_A and E_D are the acceptor and donor ionization energies, and N_D is the donor concentration. DFT predicts that the ionization energies of the carbon acceptor and nitrogen vacancy donor are 1.9 and 1.5 eV, respectively. In addition it also predicts that a maximum concentration of 1×10^{19} cm^{-3} of ionized nitrogen vacancies are present at carbon concentrations of 2×10^{19} cm^{-3}. Using Equation 2, the predicted DAP energy is approximately 2.7 eV, which matches well with the experimentally observed emission at 2.8 eV. Figure 5 shows the predicted and observed transitions between the C_N^- state, the V_N^+ state and conduction band minimum (CBM) including Stokes shifts (23).

In more detail, the transition energy of the DAP is predicted to be about 2.6 eV when accounting for the Stokes shifts and relaxation, similar to the prediction using Equation 2. This satisfies the model for emission through a DAP transition, but it would also require that an absorption around 2.8 eV must exist and that there must be efficient excitation of the carbon impurity via the 4.7 eV emission. The 2.6 to 2.8 eV absorption from the acceptor to the donor is not expected to be as significant as the 4.7 eV absorption band due to its necessary lower transition probability however Lu et al. has reported on an absorption band around 2.7 eV with a maximum absorption coefficient around 20 cm^{-1}

(24). This absorption gives the AlN bulk crystal the common yellowish "amber" coloration, but it would not be apparent on thin wafers or films. The requirement for a strong excitation channel through the 4.7 eV absorption band from the carbon impurity is validated through the PLE map shown in Figure 6.

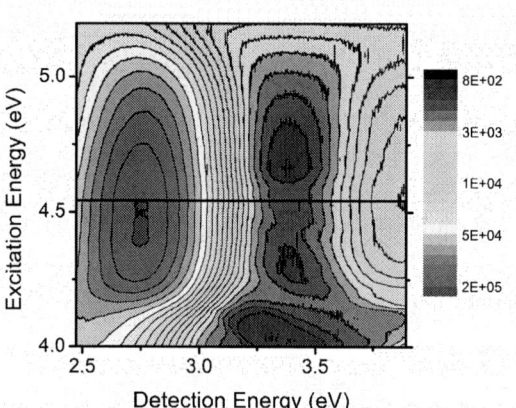

Figure 6. PLE map on an AlN single crystal substrate taken near the region of the two emissions described in Figure 5.

These measurements show a clear excitation channel around 4.4 to 4.6 eV (onset around 4.2 eV) for both the 3.9 and 2.8 eV emissions. The significant energy difference between the excitation channel between 4.4 to 4.6 eV and the emission at 2.8 eV strongly supports the identification of a donor-acceptor pair transition. Furthermore, both emissions turning on at the same energy is strong evidence that they originate from the C_N^- state as predicted by the model. Further discussion of these PLE maps will be published elsewhere.

Conclusion

The previous review discussed the problem of UV transparency in AlN crystals based on the identification of the point defects associated with the absorption and emission transitions. Based on the consistency between the DFT model predictions and experiments designed to directly address the model, we identified C_N- and V_N^+ as the point defects directly associated with the main absorption at 4.7 eV and the emission at 2.8 eV. The main absorption is due to a transition between the C_N- state and conduction band, while the emission arises from a DAP transition between the donor V_N^+ and acceptor C_N^- through the 4.7 eV excitation channel. Several other hypotheses have been discussed to explain the same absorption and emission profiles. Nevertheless, they did not consider either aspects of the process or the impurity profile of the crystals studied. Nevertheless, removal of this absorption band is desirable to make these substrates transparent and useful for deep UV optoelectronics applications. A simple way to achieve this, as discussed in the previous sections is through the removal of carbon. This task may not be easily achieved, as carbon may be an integral part of the growth process. An alternative already demonstrated is through co-doping with Si and the formation of a Si-C complex that helps remove the undesirable absorption band. Nevertheless, the formation of this complex may be just a part of the observed effect, thus research is still ongoing to elucidate possible alternatives to achieve better transparency on these substrates.

Acknowledgments

This work was supported by ARPA-E grant DE-AR0000299; I.Bryan and B. Gaddy would like to acknowledge NDSEG Fellowship under and awarded by DoD, Air Force Office of Scientific Research; D. Alden would like to acknowledge CONACYT-Mexico for their financial support and a grant of computer time from the DoD High Performance Computing Modernization Program at US AFRL, US AERDC, and Navy DoD SRC.

Bulk AlN substrates used in this study were provided by HexaTech Inc. Preparation and polishing of the HVPE samples was realized by T. Kinoshita and T. Nagashima in Tokuyama Corp.

References

1. Kneissl, M., Z. Yang, M. Teepe, C. Knollenberg, O. Schmidt, P. Kiesel, N.M. Johnson, S. Schujman, and L.J. Schowalter, *Journal of Applied Physics*, **101**, 123103-5 (2007).
2. Kinoshita, T., T. Nagashima, T. Obata, S. Takashima, R. Yamamoto, R. Togashi, Y. Kumagai, R. Schlesser, R. Collazo, and A. Koukitu, *Applied Physics Express*, **8**, 061003 (2015).
3. Baliga, B.J., *Semiconductor Science and Technology*, **28**, 074011 (2013).
4. Dalmau, R., B. Moody, R. Schlesser, S. Mita, J. Xie, M. Feneberg, B. Neuschl, K. Thonke, R. Collazo, A. Rice, J. Tweedie, and Z. Sitar, *Journal of The Electrochemical Society*, **158**, H530-H535 (2011).
5. Kinoshita, T., T. Obata, T. Nagashima, H. Yanagi, B. Moody, S. Mita, S.-i. Inoue, Y. Kumagai, A. Koukitu, and Z. Sitar, *Applied Physics Express*, **6**, 092103 (2013).
6. Inoue, S.-i., T. Naoki, T. Kinoshita, T. Obata, and H. Yanagi, *Applied Physics Letters*, **106**, 131104 (2015).
7. Xie, J., S. Mita, Z. Bryan, W. Guo, L. Hussey, B. Moody, R. Schlesser, R. Kirste, M. Gerhold, R. Collazo, and Z. Sitar, *Applied Physics Letters*, **102**, 171102-4 (2013).
8. Guo, W., Z. Bryan, J.Q. Xie, R. Kirste, S. Mita, I. Bryan, L. Hussey, M. Bobea, B. Haidet, M. Gerhold, R. Collazo, and Z. Sitar, *Journal of Applied Physics*, **115**, 103108 (2014).
9. Strassburg, M., J. Senawiratne, N. Dietz, U. Haboeck, A. Hoffmann, V. Noveski, R. Dalmau, R. Schlesser, and Z. Sitar, *Journal of Applied Physics*, **96**, 5870-5876 (2004).
10. Bickermann, M., P. Heimann, and B.M. Epelbaum, *physica status solidi (c)*, **3**, 1902-1906 (2006).
11. Bickermann, M., B.M. Epelbaum, O. Filip, P. Heimann, S. Nagata, and A. Winnacker, *physica status solidi (c)*, **7**, 21-24 (2010).
12. Collazo, R., J. Xie, B.E. Gaddy, Z. Bryan, R. Kirste, M. Hoffmann, R. Dalmau, B. Moody, Y. Kumagai, T. Nagashima, Y. Kubota, T. Kinoshita, A. Koukitu, D.L. Irving, and Z. Sitar, *Applied Physics Letters*, **100**, 191914-5 (2012).

13. Bickermann, M., B.M. Epelbaum, O. Filip, B. Tautz, P. Heimann, and A. Winnacker, *physica status solidi (c)*, **9,** 449-452 (2012).

14. Schulz, T., M. Albrecht, K. Irmscher, C. Hartmann, J. Wollweber, and R. Fornari, *physica status solidi (b)*, **248,** 1513-1518 (2011).

15. Slack, G.A., L.J. Schowalter, D. Morelli, and J.A. Freitas, *Journal of Crystal Growth*, **246,** 287-298 (2002).

16. Herro, Z.G., D. Zhuang, R. Schlesser, and Z. Sitar, *Journal of Crystal Growth*, **312,** 2519-2521 (2010).

17. Lu, P., R. Collazo, R.F. Dalmau, G. Durkaya, N. Dietz, B. Raghothamachar, M. Dudley, and Z. Sitar, *Journal of Crystal Growth*, **312,** 58-63 (2009).

18. Herro, Z.G., D. Zhuang, R. Schlesser, R. Collazo, and Z. Sitar, *Journal of Crystal Growth*, **286,** 205-208 (2006).

19. Kumagai, Y., Y. Enatsu, M. Ishizuki, Y. Kubota, J. Tajima, T. Nagashima, H. Murakami, K. Takada, and A. Koukitu, *Journal of Crystal Growth*, **312,** 2530-2536 (2010).

20. Kumagai, Y., Y. Kubota, T. Nagashima, T. Kinoshita, R. Dalmau, R. Schlesser, B. Moody, J.Q. Xie, H. Murakami, A. Koukitu, and Z. Sitar, *Applied Physics Express*, **5,** (2012).

21. Gaddy, B.E., Z. Bryan, I. Bryan, R. Kirste, J. Xie, R. Dalmau, B. Moody, Y. Kumagai, T. Nagashima, Y. Kubota, T. Kinoshita, A. Koukitu, Z. Sitar, R. Collazo, and D.L. Irving, *Applied Physics Letters*, **103,** 161901 (2013).

22. Nagashima, T., Y. Kubota, T. Kinoshita, Y. Kumagai, J. Xie, R. Collazo, H. Murakami, H. Okamoto, A. Koukitu, and Z. Sitar, *Applied Physics Express*, **5,** 125501 (2012).

23. Gaddy, B.E., Z. Bryan, I. Bryan, J. Xie, R. Dalmau, B. Moody, Y. Kumagai, T. Nagashima, Y. Kubota, T. Kinoshita, A. Koukitu, R. Kirste, Z. Sitar, R. Collazo, and D.L. Irving, *Applied Physics Letters*, **104,** 202106 (2014).

24. Lu, P., R. Collazo, R.F. Dalmau, G. Durkaya, N. Dietz, and Z. Sitar, *Applied Physics Letters*, **93,** 131922-3 (2008).

Growth of GaN/InGaN Films and Heterostructures Via Super-Atmospheric MOCVD

J. R. Krause[a], E. B. Stokes[a]

[a] Optical Science and Engineering, UNC Charlotte, Charlotte, North Carolina 28262, USA

In the interest of improving crystalline quality and optical performance of MOCVD grown semiconductors a unique super-atmospheric reactor was designed and fabricated. This reactor has since been used to fabricate GaN/InGaN multi-quantum-well heterostructures under superatmospheric growth conditions. The resulting samples were analyzed through in-situ and ex-situ measurements.

Introduction

Metal-Organic-Chemical-Vapor-Deposition (MOCVD) has become the leading method of industrial III-V semiconductor growth in the modern world. Particularly its effectiveness in the growth of optoelectronic device structures has fueled a revolution in the field of solid-state lighting. III-nitride material systems such as Gallium-Nitride (GaN), Indium-Nitride (InN) and Gallium-Arsenide (GaAs) have been of particular interest due to possessing intrinsic band-gaps in and around the visible wavelengths.

Modern MOCVD Reactors isolate the chemical reactions required for epitaxy from the outside world and limit the introduction of undesired elements into the crystal being grown. Typically available reactors are designed to operate at sub-atmospheric pressure, less than 760 torr.

Experimental evidence has indicated that MOCVD growth performed at super-atmospheric reactor pressures (greater than 760 torr) may have benefits[1,2].
From a theoretical perspective, thermodynamics indicate that elevated growth pressures may help to address the issue of the Indium adatom desorption during InGaN growth[3,4].

To test these theories a superatmospheric MOCVD reactor was designed and assembled at the University of North Carolina at Charlotte. This system features a number of refinements intended to promote high-pressure MOCVD research including mechanisms to limit gas-phase reactions and promote predictable precursor flow characteristics[5].

The body of this work observes the effects of modified V/III ratio, sample rotational speed and heterostructure dimensions while operating under super-atmospheric conditions. In particular, temperature dependence of photo luminescent output and in-situ reflectometry are analyzed. Additional modifications to the growth recipe and their effects are also addressed.

Experimental

Reactor All experiments were carried out on the experimental High-Pressure MOCVD reactor located in UNC-Charlotte. During growth group-V species were supplied by high-purity Ammonia while group-III species were supplied by Trimethylindium(TMIn) and/or Triethylgallium(TEGa). Nitrogen was used as a carrier gas.

Structure Epitaxial layers and multiple quantum well (MQW) heterostructures were grown on templates typically composed of 2 um of GaN on sapphire. Intended structures consisted of a 200nm layer of GaN followed by 4 periods of Quantum well/Barrier followed by an additional 20nm nominal cap layer of GaN.

Recipe Each growth was accomplished through an automated recipe. The recipe used in the experiment begins with a thermal cleaning with NH3 atmosphere followed by a low temperature nucleation layer. A high-temperature GaN layer is then deposited before beginning the 4 periods of Quantum Well/barrier structure. Throughout all experiments the reactor pressure was maintained within 0.1 atmospheres of 2.6 atm absolute pressure. The specifics of V/III ratio and heterostructure deposition times were modified for experimental purposes.

Photoluminescence Measurements

V/III Ratio Observation of growth rate as a function of V/III ratio and molecular flux of precursors gives insight into the whether the growth rate is being limited by lack of group-V or group-III species. Furthermore, modification of substrate temperature can offer information regarding the existence of a Mass-Transport limited or Reaction limited growth regime. Table 1 outlines the growth parameters used to address these questions.

TABLE I. V/III Ratio experimental setup

Sample ID	QW Temp (C)	V/III Ratio	Group-III Flux (umol/min)	Growth Rate (nm/min)
A	930	41500	6.3	0.75
B	930	21000	6.3	0.9
C	930	16000	16.2	2.2
D	850	16000	16.2	2.2

Sample Rotation Speed The kinetics of process gases as they come into proximity with the substrate are expected to be strongly influenced by the susceptor rotation speed. CFD analysis of the reactor geometry predicts an increase in incident velocity of process gases with increasing rotation speed. To investigate this prediction four samples were grown with susceptor rotational rates of 100, 200, 300 and 400 rpm.

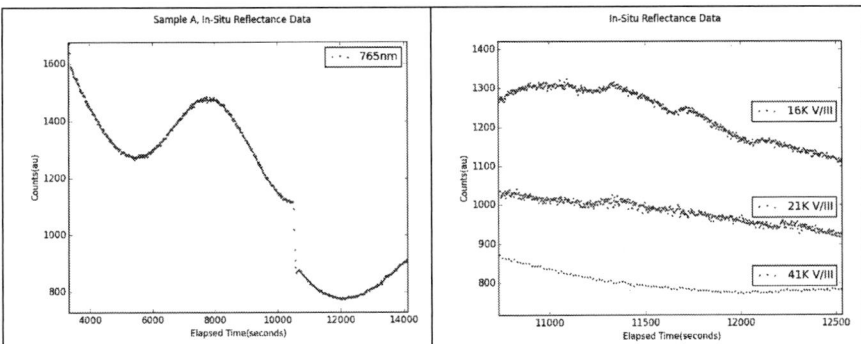

Figure 1. (Left) In-Situ Reflectance data captured during sample A growth showing oscillations typical of increasing film thickness captured at 765 nm. (Right) A selection of In-Situ data from Samples A,B and C showing a close-up of the multiple-quantum-well growth step. Transient increases in reflectance are visible during the deposition of the InGaN quantum wells.

Results and Discussion

In-situ reflectometry reveals periodic spikes in sample reflectance during heterostructure growth, Figure 1. The timing of the rising edge of these spikes corresponds to the addition of TMIn into the reactor for quantum-well growth. These reflectance spikes tend to increase in amplitude with decreasing V/III ratio. A plausible explanation is that in the absence of sufficient group-V species there is a greater accumulation of un-incorporated metallic adatoms on the sample surface. This accumulation could be further exacerbated by the elevated growth pressure that would suppress the rate of desorption.

The Growth Rate column in Table 1 is estimated from sinusoidal curve fits from in-situ data captured during each growth. From Sample A to B a small increase in growth-rate is observed despite a lack of increase in Group-III molecular flux. This result can be understood as a suppression of desorption facilitated by a decrease in abundance of reactive hydrogen species generated by cracking of Ammonia molecules.

Samples A and C both feature the same Group-V molar flux. We observe that the ratio of the growth rates is within 15% of the ratio of Group-III molar flux. This implies that growth rate is being primarily limited by availability of Group-III adatoms.

This theory is further supported by the similarity in growth rates between samples C and D, where sample D was produced with a lower substrate temperature. Substrate temperature is a major factor in cracking-efficiency of Ammonia yet the expected decrease in Group-V availability does not measurably decrease growth rate. However, this result could be confounded by a decreased desorption of Group-III adatoms that would be expected at lower temperatures.

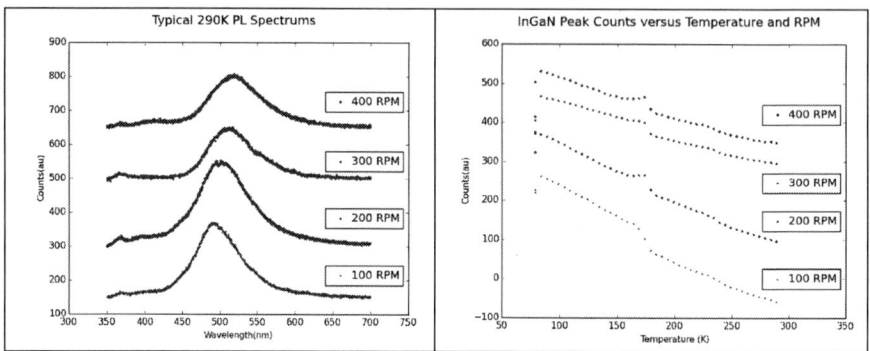

FIGURE 2. (Left) Typical 290K Photoluminescence spectrums for 100, 200, 300 and 400 RPM samples. (Right) Brightness of InGaN PL emission peak versus temperature at 100, 200, 300, and 400 RPM.

TABLE II. RPM Experiment results summary

Sample RPM	InGaN Peak @ 290K (nm)	Estimated IQE (Counts@300K/Counts@80K)*0.8
100	495	34
200	503	37.5
300	513	44
400	519	39

Variation of susceptor rotational speed most notably impacts the InGaN emission peak location as seen in Figure 2 and Table II. Peak wavelength and susceptor rotational speed seem to vary directly with increasing speeds producing a red shift of emission peak location. An enhanced growth rate at higher RPMs offers two possibilities for explaining this result.

First, in all samples the heterostructure was grown with equivalent timings resulting in the resulting quantum well thickness being largely governed by growth rate. A higher growth rate would then contribute to thicker InGaN quantum wells that would be expected to have a red-shifted emission relative to a thinner quantum well of equivalent band-gap. Second, enhanced growth rate is expected to increase incorporation of indium species into the crystal lattice since the indium adatoms have relatively less time to desorb from the sample surface before being incorporated[6].

Further investigation of the effect on growth rate will need to be carried out by extended 'calibration' growths in which a large number of in-situ reflectance oscillations will be captured. With more in-situ data the growth rate can be calculated more accurately presenting the opportunity for more in-depth analysis of the origin of the emission peak shift.

Acknowledgements

The authors would like to thank the National Science Foundation (Award # 0821590) for funding the design and creation of the superatmospheric MOCVD system used for these experiments. Furthermore we would like to thank Kyma Technologies for supplying GaN templates and expertise.

References

1. T. Takeuchi, Y.-L. Chang, A. Tandon, D. Bour, S. Corzine, R. Twist, M. Tan, and H.-C. Luan; Appl. Phys. Lett., vol. 80, pg. 2445, 2002
2. J. McChesney, P. M. Bridenbaugh, and P. B. O'Connor; Mater. Res. Bull., vol. 5, pg. 783, 1970
3. Durkaya, Goksel, et al. "Growth temperature-phase stability relation in In 1-x Ga x N epilayers grown by high-pressure CVD." *MRS Proceedings*. Vol. 1202. Cambridge University Press, 2009.
4. Schenk, H. P. D., et al. "Indium incorporation above 800 C during metalorganic vapor phase epitaxy of InGaN." *Applied physics letters* 75.17 (1999): 2587-2589.
5. Melton, Andrew G., et al. "Superatmospheric MOCVD reactor design for high quality InGaN growth." *ECS Transactions* 45.7 (2012): 73-77.
6. Keller, S., et al. "Growth and characterization of bulk InGaN films and quantum wells." *Applied Physics Letters* 68 (1996): 3147-3149.

46

New Directions in GaN Photonics enabled by Electrochemical Processes

C. Zhang[a], G. Yuan[a], K. Xiong[a], S. H. Park and J. Han[a]

[a] Department of Electrical Engineering, Yale University, New Haven, Connecticut 06511, USA

> We have developed a novel conductivity based selective electrochemical etching to introduce nanometer sized pores into GaN. The nanoporous (NP) GaN can be considered as a new form of GaN with an unprecedented tunability in optical index. The advantages of NP-GaN for both edge-emitting laser diodes and vertical surface-emitting laser diodes (VCSEL) are subsequently exhibited.

Introduction

The purpose of this paper is to demonstrate the possibility and potential of using nanoporous (NP) GaN to provide unprecedented tunability in the index of refraction while maintaining perfectly lattice matched to GaN. NP-GaN also presents a nearly lossless character from scattering and excellent electrical conductivity. We first introduce our conductivity based electro-chemical (EC) etching procedure to fabricate nanoporous GaN and its optical and electrical properties. In the following two sections, we will give two examples of optical engineering by using NP-GaN. First, the modal gain in an all nitride waveguide is engineered and the lasing threshold of an edge-emitting LD under optical pumping is reduced by two times as the result of the optical engineering. Second, we demonstrate record high reflectance (R > 99.5 %) from epitaxial GaN DBRs and a low threshold VCSEL by optical pumping.

Nanoporous GaN

The Electrochemical etching of GaN

The process of porosifying GaN electrochemically (EC), discovered recently, will be briefly discussed here (1,2). When a positive anodic bias is applied to an n^+-type GaN sample immersed in an acid-based electrolyte, the n^+-GaN becomes oxidized by holes at the surface inversion layer. The surface oxide layer is subsequently dissolved in suitable electrolyte.[3] With the use of EC etching, one could either etch or porosify GaN according to specific process conditions (2–5). In an EC etching experiment, the two most important parameters are the anodic bias and conductivity of the layers. Without loss of generality, we show in Figure 1a an etching phase diagram, which can be divided into three regions. When the applied bias or the doping concentration is low, no chemical reactions occur and GaN remains intact (yellow region). As the applied bias or the doping concentration increases, electrostatic breakdown occurs with the injection of holes to certain localized hot spots, resulting in the formation of porous structures through localized dissolution (blue region). At an even higher applied bias or with higher doping concentration, electro-polishing (complete etching) takes place (purple region).

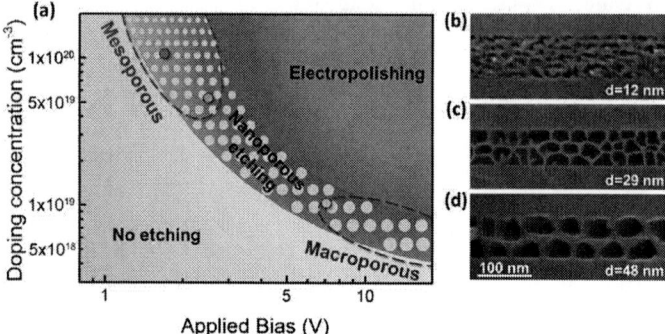

Figure 1. (a) A processing phase diagram for EC etching. The red dashed line encircles mesoporous region for photonic applications, and the blue dashed line encompasses macroporous region for light scattering applications. (b-d) SEM images of NP-GaN etched according to the red (b), green (c) and yellow (d) conditions in (a).

<u>The Optical and Electrical Properties of NP-GaN</u>

Propagation of electromagnetic waves in a heterogeneous, non-absorbing medium consisting of small air voids has been analyzed (6). When the pore diameter (d) is much less than the wavelength, specifically when the scattering factor $\chi = \pi d/\lambda < 0.2$, it was established through numerical modeling that, the index of refraction of a nanoporous medium can be described by the volume average theory (VAT):

$$n_{eff} = sqrt\,[(1- \varphi)\, n_{GaN}^2 + \varphi\, n_{air}^2] \qquad\qquad [1]$$

where φ is the porosity.[6] For a photonic device operating in the blue or green wavelength regime, the nano-pores or micro-pores will satisfy this criterion. NP-GaN therefore offers an elegant solution as a low-index medium to provide optical confinement to GaN while maintaining an aluminum-free, all-GaN structure that can be easily prepared (homo)epitaxially.

We also studied the transport property of porous GaN using Hall measurement. An n^+-GaN layer ($n > 1 \times 10^{20}$ cm^{-3}) was porosified and the measured concentration and mobility were plotted versus the porosity (Figure 2). While the effective carrier concentration is reduced as the porosity increases, the overall carrier concentration still exceeds 1×10^{18} cm^{-3} for a porosity of up to 55 %, and the mobility remains nearly constant at 100 cm^2/V-sec.

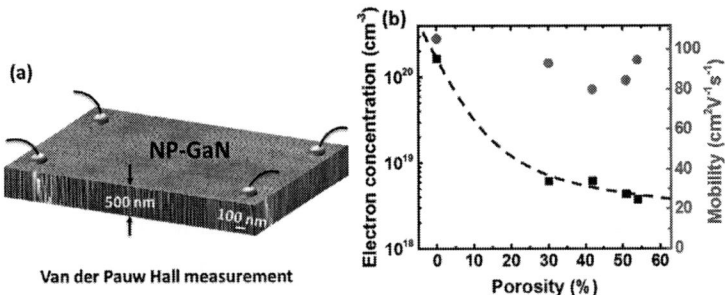

Figure 2. (a). Schematic configuration of Van der Pauw Hall measurement of nanoporous GaN. (b). The analysis of electron concentration and mobility with respect to porosity in nanoporous GaN.

Engineering the Modal Gain in edge-emitting LD (7)

Optical Confinement Factor and Modal Gain

In an optical waveguide, which is the essential part of an edge-emitting LD, the modal gain (Γg_{mat}) is defined as the product of the material gain g_{mat} (from the population-inverted quantum wells) and the optical confinement factor Γ (8,9). Γ is defined by the fraction of the optical transverse mode that resides in the quantum well region (where population inversion takes place). A low Γ implies that most of the coherent photons are outside the quantum well regions (in barriers, waveguiding, or even cladding layers) and cannot interact with injected carriers for stimulated emission. The index contrast between the cladding layer and the core layer Δn and the thickness of the core layer d_{core} critically determine the waveguide's Γ, as plotted in Figure 3. For the small index contrast ($\Delta n < 0.05$) state of the art (SOTA) laser diodes with AlGaN cladding, the Γ is capped at ~ 3% (blue color) no matter how the d_{core} is optimized. On the other hand, by replacing the AlGaN cladding layer with NP-GaN ($\Delta n > 0.4$), the optical confinement can be increased to > 9% (red color). In addition, with such large Δn, variation of d_{core} offers a tunability of Γ in a range of 4% to 9%.

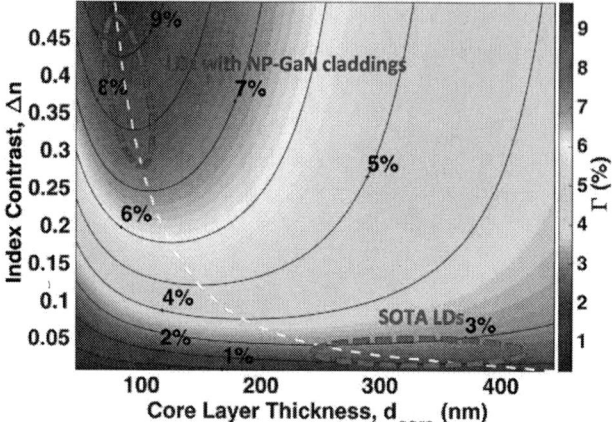

Figure 3. False-color contour plot of Γ as a function of Δn and d_{core} (the high index region sandwiched between two NP-GaN layers). Contour curves with identical Γ at integer percentage are denoted with solid black lines and numerically labeled. The white dashed line in Figure 3 indicates the optimal combination of Δn and d_{core} to achieve specific Γ maxima.

Since the modal gain is the product of the optical confinement factor and the material gain, such tunability of the Γ provides us a unique opportunity to experimentally study the increase of modal gain and the reduction of lasing threshold.

<u>Fabrication of Strongly-confined Laser Cavity with High Δn</u>

The process flow for the fabrication of high Δn edge-emitting laser cavity for optical pumping is illustrated schematically in Figure 4. The samples were first lithographically patterned and then dry etched to expose the extremely high n$^+$-doping layer. The n$^+$-doping layer was subsequently electrochemically etched into nanoporous GaN as the bottom cladding layer based on the etching phase diagram (Figure 1) in the first section. In this proof of concept demonstration, we used air as the "provisional" upper cladding to avoid unintentional absorption from the top cladding layer. After the electrochemical etching, samples were then cleaved along the GaN m-plane to form end facets.

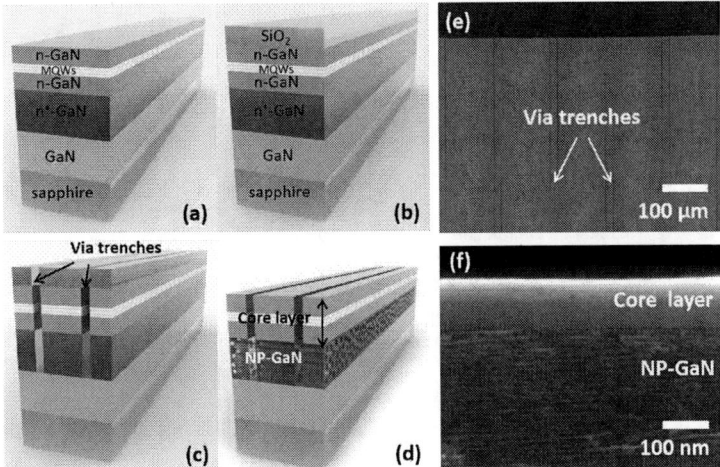

Figure 4 Schematics of the fabrication processes: (a) as grown, (b) SiO₂ capping, (c) Via trenches formation by RIE, and (d) nanoporous formation by EC etching. (e) Nomarski image of the waveguides. (f) Cross-sectional SEM of the waveguide. (The top oxide cap is etched by BOE before optical excitation.)

<u>Study of the Modal Gain and Lasing Threshold by Optical Pumping</u>

We adopted the variable stripe length (VSL) method to experimentally extract the net modal gain ($G_{net} = \Gamma g_{mat} - \alpha$) (10–13). Samples were optically pumped by a cylindrical lens focused laser beam with a stripe-like geometry. The width of the stripe was fixed at 10 μm while the length of the optically pumped region was controlled by an actuated slit. Amplified spontaneous emission (ASE) emitting from the edge of the sample was measured as a function of the excitation length L. The measured intensity I_{ASE} should depend exponentially on the excitation length L through the equation (10,11):

$$I_{ASE}(L) = (I_{spon}A/G_{net})[\exp(G_{net}L) - 1]$$ [2]

where I_{spon} is the spontaneous emission rate per unit volume, A is the cross-sectional area, and G_{net} the net modal gain due to stimulated emission. Based on this model, a direct proof of ASE is evidenced by an exponential dependence of the output light intensity on the length of the stripe excitation (L), which is shown in Figure 5a. The pumping power density was kept constant at 1.07 MW/cm², and the linear regions of the semi-logarithmic plot indicate exponential increase of light intensity I_{ASE} as the excitation length L increases, confirming ASE. Curve fitting of the linear regions (colored solid lines) gives a measurement of the net modal gain G_{net} to be 32.7, 46.0, and 70.6 cm⁻¹ for the three samples with a Γ of 4%, 6%, and 9%, respectively.

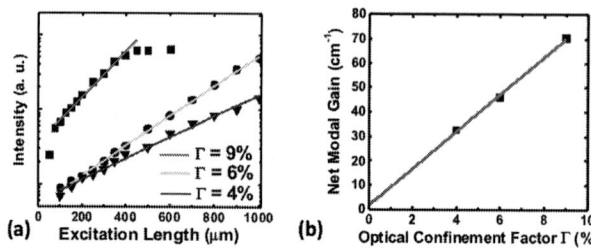

(a) Excitation Length (μm)

(b) Optical Confinement Factor Γ (%)

Figure 5 (a) Amplified spontaneous emission (ASE) intensity as a function of excitation length, with different Γ waveguides under the same pump power. (b) Extracted net modal gain as a function of Γ.

The linear correlation between Γ and G_{net} is clearly observed in Figure 5b. From the slope, the material gain g_{mat} is calculated to be ~ 760 cm^{-1} under the very pumping level. The VSL method thus proves that modal gain ($G_{modal} = \Gamma g_{mat}$) can be engineered and enhanced by more than 100% using an essentially all-GaN core-cladding design. An atitonal note needs to be made for samples with a high confinement (for instance Γ = 9%): output saturation of I_{ASE} with longer excitation length ($L > 400$ μm) was due to the finite supply of carriers.

The threshold of a laser cavity was also investigated by using the same stipe-like geometry optical pumping. An increase of the optical confinement, thus modal gain, would decrease the excitation power density (P_{th}) required to reach the lasing threshold. Figure 6a shows the threshold behavior of the three laser diodes under optical pumping, with different Γ of 4%, 6%, and 9%. Lasing threshold P_{th} was ~ 1.2, 0.9, and 0.6 MW/cm^2 for the Γ = 4%, 6%, and 9% sample, respectively. A reduction of lasing threshold by a factor of two was therefore achieved by increasing the Γ from 4% to 9%.

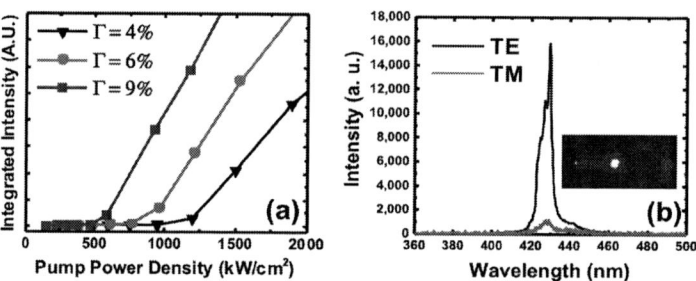

Figure 6 (a) Threshold behavior of three cavities of Γ = 4%, 6%, and 9%. With increase of the Γ, the reduction of P_{th} is clear. (b) The emission spectrum after a linear polarizer: TE/TM > 10, which again confirms lasing behavior. The inset is a microscopic photo of an emitting cavity.

High reflective DBR for VCSEL(14)

The fabrication of NP-GaN DBR

Using nanoporous GaN medium, we prepared a DBR made of GaN/nanoporous GaN. Figure 7a-c illustrates the process flow of making the nanoporous GaN DBR mirrors. First an epitaxial structure consisting of alternating n^+-GaN/GaN layers is grown (Figure 7a). Since the EC etching is designed in this case to proceed laterally, the sample needs to be lithographically patterned with trenches (via windows) to expose the sidewalls of the alternating layers (Figure 7b). The sample surface is covered by SiO_2, while the edge (not shown) is connected to a source meter so the anodic bias can be applied. EC etching is then conducted where lateral porosification proceeds from the exposed sidewalls in the direction perpendicular to the sidewall surface to form parallel nanopores (Figure 7c).

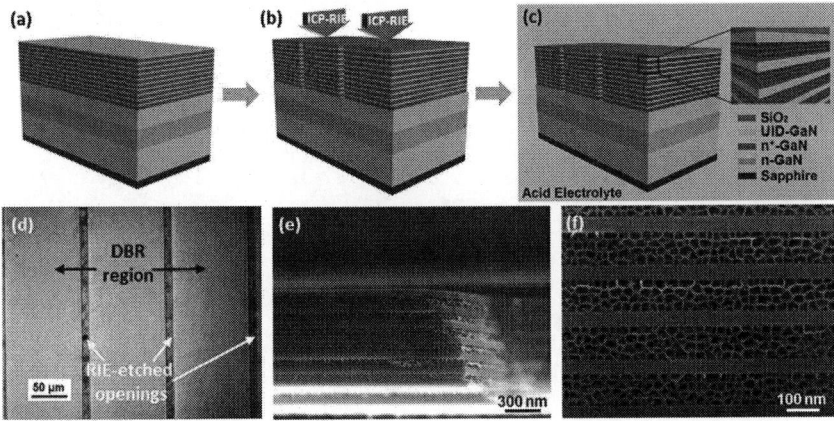

Figure 7. (a-c) A schematic process flow of forming highly reflective DBR mirror with NP-GaN layers. (a) Epitaxial growth of $\lambda/4$ n^+-GaN/GaN structures, (b) opening via windows through dry etching to expose the sidewalls of n^+-GaN layers and (c) performing electrochemical etching to convert n^+-GaN into NP-GaN laterally and selectively. The NP-GaN bears a morphology of parallel nanotubes (inset). (d) Top-view optical micrograph showing via openings (dark green regions labeled by white arrows) and the highly reflective GaN/NP-GaN DBR regions (light green regions labeled by black arrows). (e) A cross-sectional SEM image of a GaN/NP-GaN DBR structure. (f) A zoomed-in cross-sectional SEM image of a GaN/NP-GaN DBR structure.

Figure 7d is a top-view optical microscope image of such a sample that has been patterned and laterally porosified. The dark green patterns are vias/trenches created by inductively coupled plasma reactive-ion etching (ICP-RIE). Around these patterns nanoporous GaN is formed laterally, and these NP-GaN regions appear as light green color under the optical microscope due to a greatly increased reflectance. The rate of lateral porosification is ~ 5 μm/min. The area of the DBR region easily exceeds 50 μm, which is sufficiently large for the fabrication of vertical cavity surface emitting lasers (VCSELs). Figure 7e shows a cross-sectional scanning electron microscopy (SEM) image of a porous GaN DBR near the center of the cross-shaped alignment mark. A closeup SEM image of the same structure is shown in Figure 7f with the NP-GaN having a porosity of ~ 70 % and an average pore size of 30 nm. The DBR structure is formed by the conductivity-based selectivity in EC porosification which cannot be accomplished by any other means including photo-assisted EC process.

The reflectance tunability of NP-GaN DBR

The reflectance spectrum of the GaN/NP-GaN DBR was measured with a micro-reflectance setup calibrated against a commercial silver mirror with a spot size of 20 μm. The reflectance of sapphire was also measured and compared with published data, and it was determined that the accuracy is within 0.1 %. Figure 8a shows the reflectance trace (red line) of a GaN/NP-GaN DBR sample with a stopband centered at ~ 520 nm. COMSOL simulated reflectance is shown as the blue dotted line with the cross-sectional SEM image in Figure 7f scanned digitally as an input file for simulation. The difference between the simulation and measured results is probably due to the relatively large numerical aperture (around 0.34) used in the reflectance measurement. Nevertheless, a peak reflectance of more than 99.5 % is reproducibly obtained with a stopband of at least 70 nm (Figure 8a inset). We stress the fact that the peak reflectance reported here is among the highest reported from a III-nitride epitaxial DBR structure, and the full width at half maximum (FWHM) is nearly one order of magnitude wider (15–17).

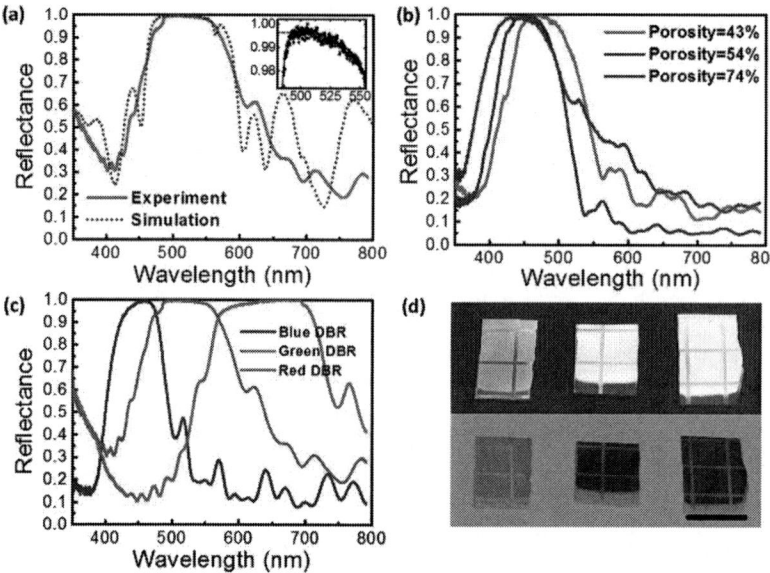

Figure 8. (a) Experimentally measured (red line) and COMSOL simulated (blue line) reflectance spectra from a DBR region. Inset: close-up of a reflectance spectrum (dot: experimental, line: fitting) with a peak reflectance exceeding 99.5%. (b) Reflectance from three NP-GaN DBRs in the blue wavelengths by tuning porosity. (c) Reflectance from three NP-GaN DBRs in the blue, green, and red wavelengths. (d) Photographs of the

three NP-GaN DBR mirrors under room-light illumination, showing process uniformity (scale bar is 1 cm). The top photograph was taken under incandescent light with continuous spectrum, the color of which reflects DBRs' reflectance spectra, and the bottom photograph was taken under fluorescent light with distributed wavelengths, the color of which represents scattering of complementary wavelengths.

To demonstrate the controllability of the NP-GaN DBR, we systematically vary two parameters. First, we use the same as-grown structures but change the anodic bias to change the porosity. Varying the porosity from 40 to 75 % causes the detuning by changing the index and the Bragg condition; the peak wavelength of the stopband can vary up to 30 nm for a blue GaN/NP-GaN DBR (Figure 8b). Separately, we prepared three structures with different thicknesses but the same doping by metal organic chemical vapor deposition. We have demonstrated highly reflective (> 99.5 %) DBR mirrors in the blue (440 nm), green (520 nm), and red (600 nm) wavelength range (Figure 8c). Since there is no heteroepitaxy involved, the preparation of these three structures are quite straightforward. The corresponding photographs of these NP-GaN DBR mirrors under room-light illumination are shown in Figure 8d.

VCSEL with NP-GaN DBR

A planar InGaN microcavity was constructed using the mesoporous GaN structure described above as the bottom DBR mirror. The microcavity for optical pumping was completed with 10 $In_{0.15}Ga_{0.85}N$ (3 nm)/GaN (8 nm) quantum wells (λ_{PL} = 450 nm) capped with 12 pairs of dielectric SiO_2/TiO_2 layers as the top mirror. The cavity was designed to have an overall effective length of 3 λ and the InGaN MQW was placed at an antinode. Optical pumping was performed with a 355 nm pulsed laser (pulse duration is 0.5 ns and pulse rate is 1 kHz). Figure 9a shows the output intensity and linewidth as functions of the excitation power with a clear lasing threshold observed at the power density of 1.5 kW/cm². Figure 9b shows the output spectra at different excitation powers (below, at and above threshold). Single mode lasing was achieved at 445 nm with a linewidth of 0.17 nm. In this preliminary design where the cavity resonance is yet to match precisely with the MQW peak emission, we measured a lasing quality (Q) factor of more than 3,000 and a below-threshold (cold cavity) Q-factor of around 500. The presence of surface-emitting laser action was further established from a far-field (40 cm away) spot pattern (inset of Figure 9b) showing a nearly circular beam profile.

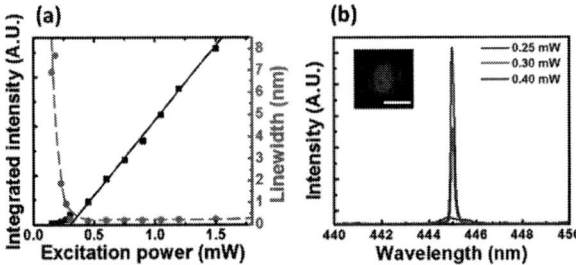

Figure 9. (a) Laser output intensity and linewidth as functions of the excitation power. (b) Output spectra at different excitation powers (below, at and above threshold) originated from an optically pumped micro-cavity with a GaN/NP-GaN bottom DBR and a dielectric top DBR. Single mode lasing at 445 nm with 0.17 nm linewidth was achieved

above threshold. Inset: far field laser spot from optical pumping of the micro-cavity (scale bar is 5 mm).

Acknowledgments

This research was supported by the National Science Foundation (NSF) under Award CMMI-1129964, and facilities used were supported by Yale SEAS cleanroom, YINQE, and NSF MRSEC DMR-1119826. We acknowledge Prof. Arto Nurmikko and Dr. Joonhee Lee at Brown University for their kind help with wafer dicing and discussions of optical measurement.

References

1. Y. Zhang et al., *Phys. Status Solidi B*, **247**, 1713–1716 (2010).
2. D. Chen, H. Xiao, and J. Han, *J. Appl. Phys.*, **112**, 064303 (2012).
3. M. J. Schwab, D. Chen, J. Han, and L. D. Pfefferle, *J. Phys. Chem. C*, **117**, 16890–16895 (2013).
4. J. Park, K. M. Song, S.-R. Jeon, J. H. Baek, and S.-W. Ryu, *Appl. Phys. Lett.*, **94**, 221907 (2009).
5. S. H. Park et al., *Nano Lett.*, **14**, 4293–4298 (2014).
6. M. M. Braun and L. Pilon, *Thin Solid Films*, **496**, 505–514 (2006).
7. S. Einfeldt et al., *J. Appl. Phys.*, **88**, 7029–7036 (2000).
8. G. P. Agrawal and N. K. Dutta, *Semiconductor Lasers*, Springer US, Boston, MA, (1995) http://link.springer.com/10.1007/978-1-4613-0481-4.
9. L. A. Coldren, S. W. Corzine, and M. L. Mashanovitch, *Diode Lasers and Photonic Integrated Circuits*, p. 740, John Wiley & Sons, (2012).
10. K. L. Shaklee and R. F. Leheny, *Appl. Phys. Lett.*, **18**, 475–477 (1971).
11. K. L. Shaklee, R. E. Nahory, and R. F. Leheny, *J. Lumin.*, **7**, 284–309 (1973).
12. L. Pavesi, L. Dal Negro, C. Mazzoleni, G. Franzò, and F. Priolo, *Nature*, **408**, 440–444 (2000).
13. S. G. Cloutier, P. A. Kossyrev, and J. Xu, *Nat. Mater.*, **4**, 887–891 (2005).
14. C. Zhang et al., *ACS Photonics*, **2**, 980–986 (2015).
15. K. E. Waldrip et al., *Appl. Phys. Lett.*, **78**, 3205–3207 (2001).
16. J. Zhang et al., *Appl. Phys. Lett.*, **80**, 3542–3544 (2002).
17. R. Butté et al., *Jpn. J. Appl. Phys.*, **44**, 7207–7216 (2005).

Chapter 4

Oxides

58

Channel scaling behavior of amorphous In-Zn-O thin film transistors with high mobility over 35 cm²/Vsec

Sunghwan Lee[a] and David C. Paine[b]

[a]Department of Mechanical Engineering, Baylor University, Waco, TX 76798, USA
[b]School of Engineering, Brown University, Providence, RI 02912, USA

We report on the channel scaling behavior of amorphous In-Zn-O (a-IZO) thin film transistor (TFT) devices. The TFTs were fabricated at room temperature where a-IZO channel and Mo metallization were deposited using dc magnetron sputtering. Both channel length (L) and width (W) of the devices were scaled down in order to maintain the aspect ratio, W/L of 20. The channel scaling of a-IZO TFTs leads to an increase in on-state drain current. However, an increase in off-state current, a decrease in field effect mobility and a negative shift in threshold voltages are also observed with decreasing channel length, known as short channel effects which are mainly attributed to an increase in the longitudinal electric field applied to channel IZO.

Introduction

Amorphous oxide semiconductors (AOSs) based on In_2O_3 are promising due mainly to high carrier mobility(1-3) and excellent optical transmittance(1) in visible regime. Therefore, AOS materials have been integrated as active layers(2, 4) and electrodes(2, 5, 6) into a variety of electronic devices such as high performance thin film transistors(2, 7) (TFTs) and fast response photo detectors(8, 9). This class of materials has been gaining particular attention in next generation high-resolution displays due to their much higher TFT field effect mobility (>10 vs 1 cm²/Vsec) (1-3) and low temperature (T) processability (RT-100 °C vs ~300 °C) (1-4) compared to conventional amorphous Si (a-Si)-based TFTs. Some AOS materials such as In-(Ga)-Zn-O is now being implemented in high performance and flexible active-matrix liquid crystal displays and active-matrix organic light emitting diodes technologies. Additional advantages of AOSs besides high mobility, low T process and optical transparency include isotropic wet etch characteristics and compatibility with mass productions all of which make this material suitable for large area, flexible, and transparent devices on inexpensive polymer substrates(10).

In previous works, we have extensively contributed to the development of high performance and stable amorphous In-Zn-O (a-IZO) TFTs. These efforts include(2, 4, 7, 11-14) the first room-temperature fabrication of high mobility, compositionally homogeneous channel/metallization a-IZO TFTs; the first measurement of the specific contact resistance in a-IZO metallized thin a-IZO channel structure; the first identification of native defect doping in a-IZO using ultra-high pressure oxidation. Amorphous/crystalline phase stability and thermal-stress-induced threshold voltage instability of a-IZO TFTs have also been reported. Metallization strategies for In_2O_3-based TFTs were proposed, and the report on the relevance of metallization selection to the performance of a-IZO TFTs validated the usefulness of the proposed strategies.

One next challenge is to scale down AOS TFTs for the implementation in ultra-high definition (UHD) displays. Since these next-generation technologies utilize much smaller pixel size in order to realize UHD resolution, AOS TFTs that are employed as pixel driving elements must be scaled down as well. When the dimensions of TFT devices (e.g., channel length and width) are reduced, the important device characteristics such as the field effect mobility, device saturation behavior and threshold voltage are likely influenced by scaling as shown in conventional MOSFET devices. However, studies on this scaling effect on the performance of AOS TFTs are currently quite limited in the literature.

In this study, we report on the effect of the scaling behaviors of a-IZO-based TFTs on the device performance. The devices with various channel lengths (L) and widths (W) were fabricated at room temperature and patterned using a photolithographical technique and a lift-off process. We have found that the TFT field effect mobility (μ_{FE}) decreases with decreasing channel length (L) from ~40 cm^2/Vsec (L=50 μm) to ~17 cm^2/Vsec (L=5 μm) in spite of the devices with the same aspect ratio. In addition, the long channel devices present excellent drain current (I_D) saturation while the short channel a-IZO TFTs lose the I_D saturation behavior and show much higher off-sate current compared to long channel TFTs. Short channel effects including V_T increase and μ_{FE} decrease of a-IZO TFTs are presented, for the first time.

Experimental details

Thin film depositions of amorphous IZO and metallization

All depositions of the channel, source/drain, and gate contact electrodes were performed at room temperature using dc magnetron sputtering with a target to substrate distance of 10 cm. Before all depositions, the sputter chamber was pumped down to a base pressure of lower than 6×10^{-6} Torr and the target was pre-sputtered for 1000 seconds to remove surface contamination and to ensure homogeneous distribution of process gas in the sputter chamber. The channel IZO films were sputtered from a commercially available 90 wt% In_2O_3-10 wt% ZnO target (Idemitsu Corp., Japan) at a deposition rate of ~0.035 nm/sec using a dc power density of 0.22 W/cm^2 at 280V, a chamber pressure of 2 mTorr, and an Ar/O_2 gas volume fraction of 86/14. For source/drain (S/D) and gate contact metallization, Mo metal films (~100 nm) were deposited at 0.88 W/cm^2 at 350 V at a rate of ~0.21 nm/sec.

a-IZO TFT device fabrication and characterizations

Bottom-gated a-IZO TFTs were fabricated on heavily-doped single crystalline n-type Si (0.003-0.005 Ωcm) substrates and thermally grown SiO_2 (50 nm-thick) was used as the gate dielectric. The channel region was masked photolithographically using S-1818 positive photoresist and wet-etched in 0.2 M HCl for 15-20 sec, to achieve on etch rate of ~1 nm/sec. The channel pads were cleaned with acetone, methanol and de-ionized water and, then, the source/drain regions were patterned using AZ-5214 lift-off photoresist prior to Mo metallization deposition. The backside of the wafer was etched with buffered hydrofluoric acid (HF) to remove any oxides, and a Mo gate contact electrode (100nm thick) was then deposited as described above. Transistor transfer and output characteristics were measured using an Agilent 4155C semiconductor parameter analyzer in the light-tight probe station.

Results and discussion

A schematic view of the bottom-gated a-IZO TFT structure is shown in Figure 1(a). In this study, devices with the channel width/channel length aspect ratio, W/L=1000/50, 500/25, 200/10, 100/5 μm/μm were fabricated in order to investigate the channel scaling behaviors of a-IZO TFTs. Transmission electron microscope (TEM) analysis was made on the resulting IZO thin films (~10 nm) and obtained images are presented in Figure 1(b). Since IZO materials are electron beam sensitive, this data was acquired using a minimum dose technique which requires the operator to focus in one area and then electronically translate to an adjacent, unexposed region, to record the image. Plan-view and associated selected area diffraction (SAD) images exhibit neither crystallographic contrast nor diffraction ring-pattern and therefore the resulting IZO films are determined to be amorphous phase.

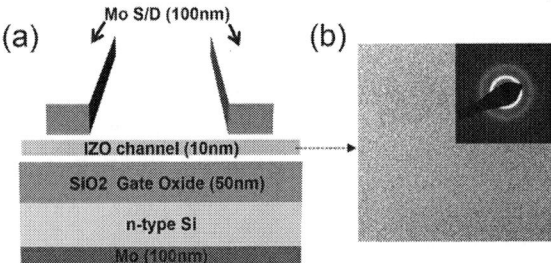

Figure 1. (a) Schematic of the bottom-gated TFT structure that utilizes a-IZO, SiO$_2$, heavily doped n-Si and Mo as channel, gate dielectric, gate and metallization materials, and (b) a plan-view TEM micrograph that presents the resulting channel IZO is in amorphous state. Inset of SAD pattern shows no crystallographic diffraction, which confirms the amorphous phase of IZO.

Figure 2 presents the TFT output characteristics of drain current (I_D) vs drain bias (V_D) as the a-IZO channel length is scaled down with scaling channel width as well so that the TFTs maintain the same aspect ratio, W/L of 20. Note that since the drain current is proportional to the channel width, the measured drain current is normalized to the channel width in order to exclude the effect of channel width on the device performance when comparing TFT characteristics. Drain current density significantly increases with scaling down the channel length from 50 to 5 μm at the same regime of V_D and V_G applied. Dashed lines are inserted at the current density of 40 μA/μm in each plot of Figure 2 (a-d), which facilitates to compare changes in drain current. For example,

I_D of the TFT with L=10 μm increases more than 2-fold than that of TFT with L=25 μm. In addition, the device off-state current increases as a result of channel scaling down.

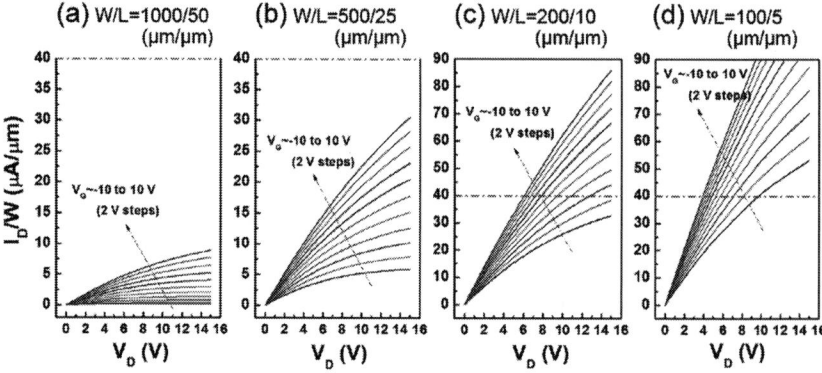

Figure 2. Typical drain current-drain bias (I_D-V_D) output characteristics of a-IZO-based TFTs at applied gate voltages in the -10 to 10 V range: the aspect ratio of (a) WL=1000/50 μm/μm, (b) WL=500/25 μm/μm, (c) WL=200/10 μm/μm and (d) WL=100/5 μm/μm.

The TFT transfer characteristics in Figure 3 were measured in the usual way by setting the drain bias (V_D=25 V in this study) and sweeping the gate bias (V_G~ –25 to 35 V). All a-IZO TFTs fabricated in this study show promisingly high on/off ratio of approximately 10^5-10^8 which is estimated by the ratio of the minimum current density (off-state) and the maximum current density (on-state) from the transfer characteristics shown in Figure 3(a). The device off-state current increases more than 2 orders of magnitude from below ~10^{-12} to ~10^{-10} A/μm with scaling down channel length of a-IZO TFTs while the on-current slightly increases. The maximum on/off drain current ratio > 10^8 is achieved from the device with W/L=500 μm /25 μm. Figure 3(b) presents plots of $(I_D)^{1/2}$ vs V_G from which the threshold voltage (V_T) of TFTs is estimated by extrapolation of the linear portion of $(I_D)^{1/2}$-V_G curves to zero drain current (x-axis). A negative shift in V_T is observed in Figure 3(b) as the channel length reduces (-7.6 V of L=50 μm vs. -16.3 V of L=5 μm). In general, negative V_T shifts occur when channel carrier density increases or trap states within the band gap release electrons, which is also associated with an increase in channel carrier density. However, since all a-IZO channels used in the present study were processed simultaneously, a reduction in TFT channel length is unlikely to yield a significant change in the resulting channel carrier density. Instead, the negative V_T shift with decreasing channel length in a-IZO TFTs is possibly understood from a decrease in the potential barrier to electrons between the source and drain which is known as drain-induced barrier lowering (DIBL)(15): as the channel length is shortened, the source and drain fields are significantly overlapped and penetrate into the middle of the channel, which lowers the potential barrier in the channel regime and leads to loss of gate control. This DIBL tremendously increases the sub-threshold current in the device and the V_T due to the weaker gate control, which is limited in longer channel devices (e.g., L=50 μm a-IZO TFTs in this study) due to un-overlapped and flat potential barrier over most parts of the device. Likewise, a reduction in channel length in a-IZO TFTs in the present study results in a negative V_T shift and an increase in off-state current. Note that the DIBL in a-IZO TFTs occurs at a channel length in the range of approximately 5-10 μm, which is much longer channel length than that (below ~1-2 μm) observed in Si-based field effect transistor devices.

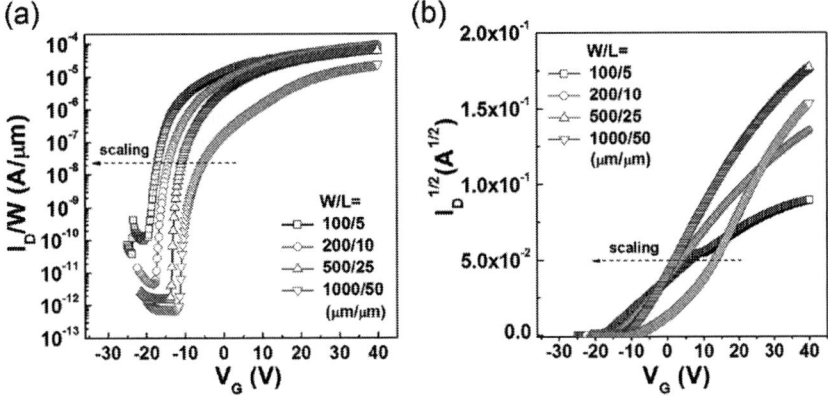

Figure 3. Transfer characteristics of (a) current density (I_D/W)-V_G and (b) $(I_D)^{1/2}$-V_G as a function of channel length from which the TFT on/off current ratio, the threshold voltages and the field effect mobility were determined using the MOSFET equation: channel scaling results in a decrease in on/off ratio (10^8 to 10^5) and μFE (~35.7 to 16.4 cm²/Vsec) and an increase in V_T (-7.6 to -16.3 V).

The TFT filed effect mobility (μFE) is determined using the MOSFET equation(16):

$$I_D = \mu_{FE} C_{ox} \frac{W}{2L} (V_G - V_T)^2$$

which relates I_D to μFE, V_G, V_T, oxide capacitance (C_{ox}=6.9×10⁻⁸ F/cm² for 50 nm-thick SiO₂), and the device aspect ratio (W/L). Promisingly high field effect mobility is achieved in these a-IZO TFTs: 16.4 cm²/Vsec (L=5 μm), 24.7 cm²/Vsec (L=10 μm), 28.9 cm²/Vsec (L=25 μm) and 35.7 cm²/Vsec (L=50 μm). Note, however, that the TFT field effect mobility significantly depends on the channel length. The largest device (W/L=1000 μm/50 μm) shows the highest field effect mobility of 35.7 cm²/Vsec while the shortest channel device (W/L=100 μm/5 μm) shows the lowest 16.4 cm²/Vsec. A decrease in *μFE* as a result of channel scaling can be understood by the velocity saturation and surface scattering in short channel devices. As the channel length is scaled down and, consequently, the longitudinal electric field (E_y) increases, the electron drift velocity ($v_d \equiv \mu_{FE}/E_y$) in the channel increases more slowly (non-linearly) with the electric field intensity than long channel TFTs (i.e., low E_y), which is known as the velocity saturation effect in short channel TFTs(15, 16). In addition, as the channel length becomes smaller, scattering at the surface and interface that occurs by the electron collision may increase and reduce the carrier mobility because the high E_y in short channel TFTs increases electron collision at the surface. Although the effective channel in bottom-gated devices forms at or near the interface between IZO and SiO_2 dielectric (i.e., similar to a buried channel field effect transistor), the utilization of thin channel IZO (~10 nm) in the devices possibly increase scattering at both the channel surface and the IZO/SiO_2 interface in the high electric field that is present in short channel IZO TFTs.

Conclusions

Amorphous IZO-based TFTs were fabricated at room-temperature using dc magnetron sputter deposition. In order to investigate the channel scaling behaviors of a-IZO-based TFTs, channel length is scaled down from 50 μm to 5 μm at a fixed aspect ratio, W/L=20. While on-current increases with decreasing IZO channel length, off-current significantly increases and the drain current is not saturated in spite of the application of drain bias of 15 V. Short-channel effects are observed: a negative shift in V_T is observed possibly due to DIBL and a decrease in μ_{FE} is attributed to an increase in scattering at the channel surface and the IZO/SiO_2 interface as a result of high electric field in short channel devices. The findings in this study may be significant to In_2O_3-based AOS TFTs including a-IZO and a-IGZO for the next-generation UHD resolution display technologies that need to implement smaller size TFTs as pixel switching elements.

Acknowledgments

The authors gratefully thank the National Science Foundation (NSF) Award No. DMR-1409590 for the financial support. S.L. acknowledges Baylor University faculty startup funds that supported this research.

References

1.   K. Nomura, H. Ohta, A. Takagi, T. Kamiya, M. Hirano and H. Hosono, *Nature*, **432**, 488 (2004).
2.   S. Lee, H. Park and D. C. Paine, *Journal of Applied Physics*, **109**, 063702 (2011).
3.   A. J. Leenheer, J. D. Perkins, M. F. A. M. van Hest, J. J. Berry, R. P. O'Hayre and D. S. Ginley, *Physical Review B*, **77** (2008).
4.   S. Lee and D. C. Paine, *Applied Physics Letters*, **104**, 252103 (2014).
5.   A. Sato, K. Abe, R. Hayashi, H. Kumomi, K. Nomura, T. Kamiya, M. Hirano and H. Hosono, *Applied Physics Letters*, **94** (2009).
6.   D. C. Paine, B. Yaglioglu, Z. Beiley and S. Lee, *Thin Solid Films*, **516**, 5894 (2008).
7.   S. Lee, H. Park and D. C. Paine, *Thin Solid Films*, **520**, 3769 (2012).

8.   S. Cosentino, P. Liu, S. T. Le, S. Lee, D. Paine, A. Zaslavsky, D. Pacifici, S. Mirabella, M. Miritello, I. Crupi and A. Terrasi, *Applied Physics Letters*, **98** (2011).

9.   L. Pei, S. Cosentino, S. T. Le, S. Lee, D. Paine, A. Zaslavsky, D. Pacifici, S. Mirabella, M. Miritello, I. Crupi and A. Terrasi, *Journal of Applied Physics*, **112**, 083103 (5 pp.) (2012).

10.   S. J. Pearton, W. Lim, E. Douglas, F. Ren, Y. W. Heo and D. P. Norton, in *Oxide-Based Materials and Devices*, F. H. Teherani, D. C. Look, C. W. Litton and D. J. Rogers Editors (2010).

11.   S. Lee, B. Bierig and D. C. Paine, *Thin Solid Films*, **520**, 3764 (2012).

12.   S. Lee and D. C. Paine, *Applied Physics Letters*, **98**, 262108 (2011).

13.   S. Lee and D. C. Paine, *Applied Physics Letters*, **102**, 052101 (2013).

14.   S. Lee, K. Park and D. C. Paine, *Journal of Materials Research*, **27**, 2299 (2012).

15.   Y. Taur and T. H. Ning, *Fundamentals of Modern VLSI Devices*, Cambridge University Press (2009).

16.   B. G. Streetman and S. K. Banerjee, *Solid State Electronic Devices*, Pearson Prentice Hall (2006).

Investigation on Alumina Passivation for Improved IGZO TFT Performance

T. Mudgal[a], N. Edwards[a], P. Ganesh[a], A. Bharadwaj[a], R. G. Manley[b], and K. D. Hirschman[a]

[a] Electrical & Microelectronic Engineering Department,
Rochester Institute of Technology, Rochester, New York 14623, USA
[b] Corning Incorporated, Science and Technology, Corning, New York 14870, USA

A study of the influence of back-channel alumina passivation on the operation of bottom-gate IGZO TFTs is presented. TFTs without any passivation material deposited typically exhibit best-case initial results. Regardless a passivation layer is required for device stability and process integration. The impact of passivation using alumina deposited via electron beam evaporation and atomic layer deposition (ALD) has been investigated. A decrease in subthreshold slope and channel mobility on certain treatment combinations is attributed to an inferior IGZO/alumina back-channel interface. A two-step passivation process has been developed which offers back-channel protection during device fabrication, and remains compatible with an oxidizing ambient anneal. Modifications in the passivation and annealing procedures and process integration details have resulted in a marked improvement in the performance of alumina passivated devices, with demonstrated resistance to aging.

Introduction

Oxide semiconductors have been extensively studied over the last decade for high performance display applications (1-3). Indium gallium zinc oxide (IGZO) in particular has been considered the dominant candidate for replacement of hydrogenated amorphous silicon (a-Si:H) as the channel material for thin-film transistors (TFTs) (4, 5). The low mobility of a-Si:H and high cost associated with low temperature polycrystalline silicon (LTPS) makes IGZO a leading choice for the display industry. Although the device characteristics have been impressive, the application of material for passivation of the IGZO back-channel interface has presented certain challenges. The deposition of back-channel passivation material tends to shift I-V characteristics towards depletion mode, which in some cases can be reversed by annealing in oxidizing ambient conditions (6-8). While TFTs with passivation material (WPM) offer improvements in stability and resistance to aging, they often have an increased sensitivity to back-channel defects that is not apparent on TFTs without passivation material (WoPM). Exposure of an unprotected back-channel to various chemistries involved in lithographic patterning and contact metallization may compromise device performance, thus process activity before application of back-channel protection should be avoided.

In this paper, the results of IGZO devices WoPM are compared with alumina passivated TFTs deposited by e-beam evaporation and atomic layer deposition (ALD). A two-step passivation method has been employed to address the effects of process-induced surface degradation. Annealing conditions have been modified to account for observed differences in oxidant transport (6). Secondary ion mass spectroscopy (SIMS) measurements have been employed to correlate the interpretation of electrical results to the IGZO film composition.

Experiment

Bottom-gate top-contact TFTs are fabricated using RF sputtered IGZO. The fabrication process steps are shown in Figure 1. In the original process#1 the passivation material (option for WPM devices) has been deposited at the end of fabrication, while in the modified two-step process#2 a thin 20 nm layer of alumina is deposited after IGZO sputtering to protect the back-channel surface, with a thicker 80 nm application at the end of fabrication. WoPM devices received an anneal treatment at 400 °C for 30 min in N_2 ambient with an extended ramp-down in air. WPM devices received a more aggressive oxidation treatment at 400 °C for 30 min in air ambient. The complete fabrication process utilizing e-beam evaporated alumina is described elsewhere (9). ALD alumina is deposited at 200 °C using trimethylaluminum (TMA) and water as precursors. SIMS analysis is used to investigate the change in composition after passivation material deposition.

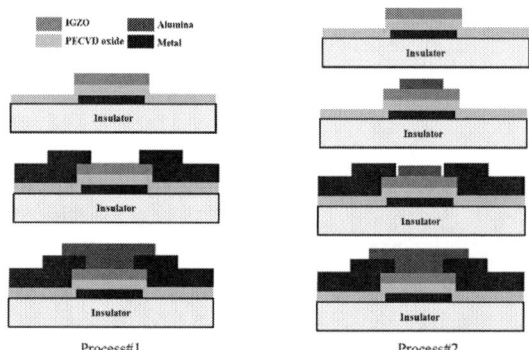

Process#1   Process#2

Figure 1. Fabrication processes for alumina-passivated IGZO TFTs. Process#1 has alumina deposition at the end, whereas process#2 has a two-step application. Both processes are followed by an oxidizing ambient anneal.

Results & Discussion

Process#1 TFTs

The transfer characteristics of a WoPM device are shown in Figure 2, with extracted operating parameters as listed. The WoPM results are the benchmark for comparison; it is imperative to have a passivation layer for device integration.

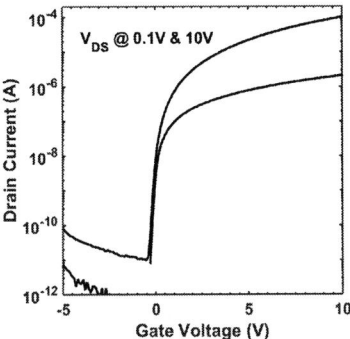

Figure 2. Transfer characteristics of a WoPM device. The TFT channel dimensions are L=24 μm & W=100 μm and the extracted parameters are: $V_T \sim 0.1$ V, $\mu_{sat} \sim 12$ cm^2/V·s & $SS \sim 130$ mV/dec.

It has been reported that ALD alumina provides a good passivation layer for ZnO TFTs (10). To explore ALD alumina as a potential passivation material for IGZO, 20 nm alumina was deposited on TFTs. These devices did not show any gate modulation (see Figure 3), with a low resistivity value of $\rho \sim 0.003$ Ω·cm. The conductivity of the IGZO channel is increased during ALD which is performed at 200 °C in vacuum. It has been shown previously that IGZO films annealed in vacuum exhibit higher conductivity due to the creation of O-vacancies (11). Also the use of water as a precursor may be responsible for hydrogen incorporation (12). Regardless of the mechanism of enhanced conductivity, it was irreversible even with the aggressive anneal treatment due to the high integrity of the thin ALD alumina layer acting as a barrier to oxidants.

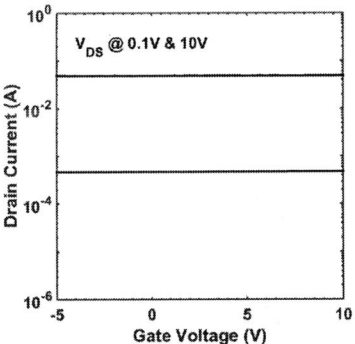

Figure 3. Process#1 WPM device with 20 nm ALD alumina, demonstrating a complete lack of channel charge modulation.

WPM devices with 100 nm evaporated alumina have been realized (9), albeit with some noted compromise in device performance. Electrical results of TFTs with 100 nm AlO_X film are shown in Figure 4. These devices show clear gate modulation, thus demonstrating successful diffusion of oxidants. The observed degradation in SS and μ_{sat} is attributed to a back-channel interface state density of $N_{IT} \sim 5 \times 10^{11} cm^{-2}$ inherent to process#1 (13).

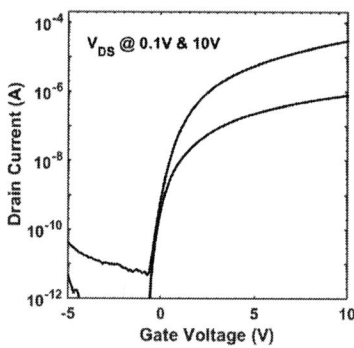

Figure 4. Process#1 WPM device with 100 nm evaporated alumina. The TFT channel dimensions were L=24 μm & W=100 μm and the extracted parameters are: $\mu_{sat} \sim 5$ cm^2/V·s & $SS \sim 280$ mV/dec.

SIMS analysis was employed to identify whether the degradation in process#1 WPM device behavior shown in Figure 4 is due to a change in composition of IGZO channel material. Measurements taken on samples before and after passivation have been published elsewhere (14), and do not support any compositional change. Degradation in performance is attributed to the IGZO/AlO_X interface due to AlO_X deposition and/or pre-passivation processes.

Process#2 TFTs

WPM TFTs were fabricated using the defined two-step passivation process#2, with transfer characteristics shown in Figure 5. These devices exhibit better μ_{sat} and SS values compared to devices fabricated using process#1. This improvement is attributed to a superior back-channel condition due to protection from chemical or physical exposure during processing. The subthreshold performance is comparable to WoPM devices. While the channel mobility appears to be lower than the WoPM device shown in Figure 2, this result may be due, in part, to process variation of physical parameters. To determine the influence of aging, WPM devices stored for more than six months in room ambient were tested, with a near-perfect overlay shown in Figure 5. A variant of process#2 that investigated the influence of an intermediate anneal following the 20 nm AlO_X application produced inferior results and was not pursued further.

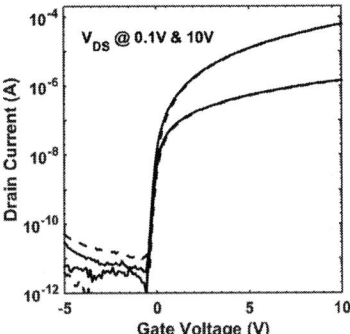

Figure 5. Process#2 WPM device with 100 nm evaporated alumina (combined 20 nm + 80 nm). A comparison of I-V characteristics taken over six months of aging demonstrates near-perfect overlay. Differences in the off-state leakage are attributed to voltage sweep conditions. The TFT channel dimensions are L=24 μm & W=100 μm and the extracted parameters are: μ_{sat} ~10 cm^2/V·s & SS ~150 mV/dec.

Conclusion

The results of using ALD alumina as a back-channel passivation material was shown to be incompatible with an oxidizing ambient anneal to establish the semiconducting properties of IGZO. WPM TFTs with evaporated AlO$_X$ deposited at the end of fabrication resulted in I-V characteristics with degradation in SS and μ_{sat} compared to WoPM devices. The two-step alumina passivation process#2 resulted in WPM TFT characteristics which compare well with WoPM device behavior, with a notable reduction in interface trap density to N_{IT} ~ 10^{11}cm^{-2}. These devices exhibit no apparent change in electrical behavior during storage in room ambient for more than six months. A single AlO$_X$ application would simplify the process sequence and is under further investigation.

Acknowledgments

The authors would like to acknowledge the support of the technical staff at the Semiconductor & Microsystems Fabrication Laboratory at RIT, and the technical staff at Corning Incorporated that provided analytical services. Financial support has been provided by Corning Incorporated and NYSTAR, through the New York State Center for Advanced Technology.

References

1. D. A. Mourey, D. L. A. Zhao and T. N. Jackson, *IEEE Electron Device Letters*, **31**, 326 (2010).
2. S. J. Seo, C. G. Choi, Y. H. Hwang and B. S. Bae, *Journal of Physics D-Applied Physics*, **42**, 035106 (2009).
3. Y. Ogo, K. Nomura, H. Yanagi, T. Kamiya, M. Hirano and H. Hosono, *Physica Status Solidi a-Applications and Materials Science*, **205**, 1920 (2008).
4. M. J. Seok, M. H. Choi, M. Mativenga, D. Geng, D. Y. Kim and J. Jang, *IEEE Electron Device Letters*, **32**, 1089 (2011).
5. T. Kamiya, K. Nomura and H. Hosono, *Science and Technology of Advanced Materials*, **11**, 044305 (2010).
6. T. Mudgal, N. Walsh, R. G. Manley and K. D. Hirschman, *MRS Online Proceedings Library*, **1692** (2014).
7. T. T. T. Nguyen, B. Aventurier, T. Terlier, J. P. Barnes and F. Templier, *Journal of Display Technology*, **11**, 554 (2015).
8. K. Nomura, T. Kamiya and H. Hosono, *Thin Solid Films*, **520**, 3778 (2012).
9. T. Mudgal, N. Walsh, R. G. Manley and K. D. Hirschman, *ECS Journal of Solid State Science and Technology*, **3**, Q3032 (2014).
10. D. A. Mourey, D. A. L. Zhao, J. Sun and T. N. Jackson, *IEEE Transactions on Electron Devices*, **57**, 530 (2010).
11. T. Mudgal, N. Walsh, R. G. Manley and K. D. Hirschman, *ECS Transactions*, **61**, 405 (2014).
12. B. Du Ahn, H. S. Shin, H. J. Kim, J. S. Park and J. K. Jeong, *Applied Physics Letters*, **93**, 203506 (2008).
13. T. Mudgal, N. Walsh, N. Edwards, R. G. Manley and K. D. Hirschman, *ECS Transactions*, **64**, 93 (2014).
14. T. Mudgal, N. Walsh, N. Edwards, R. G. Manley and K. D. Hirschman, *MRS Online Proceedings Library Archive*, **1731** (2015).

Ultra-Small ZnO Nanoparticles for Charge Storage in MOS-Memory Devices

N. El-Atab, A. Nayfeh

Institute Center for Microsystems – iMicro, EECS, Masdar Institute of Science and Technology Abu Dhabi, United Arab Emirates

ZnO-nanoparticles have gained considerable interest by industry and research due to their excellent properties. However, the agglomeration of nanoparticles is considered to be a limiting factor since it can affect the desirable physical and electronic properties of the nanoparticles. In this work, 1-to-5-nm-thick ZnO-nanoparticles deposited by dip-coating are studied. The results show that dip-coating leads to 1-D quantum confinement in ZnO (2-D nanostructures). Memory devices with ZnO-nanoparticles charge trapping layer show that a large memory window can be obtained at low operating-voltages due to the large available charge trap states in ZnO. Moreover, the excellent retention and endurance characteristics show that ZnO nanoparticles are promising for low-power memory applications.

Introduction

Recently, ZnO has gained a growing attention due to its wide-ranging applicable properties such as high electron mobility, high transparency, piezoelectricity, large direct band gap around 3.4 eV, in addition to luminescence due to its large exciton binding energy (around 60 meV) [1, 2]. Moreover, ZnO nanostructures can have tunable properties due to quantum confinement effects, in fact, ZnO is one of the few oxides that shows quantum confinement effects in an experimentally accessible size range [3], which make them promising in different fields of nanotechnology [4-9]. In specific, ZnO nanoparticles offer considerable potential as starting material for applications such as varistors, transparent conductive oxide, and for other purposes such as transparent UV-protection films and chemical sensors [10-11]. In this work, the effect of embedding dip-coated ZnO nanoparticles on their physical and electronic properties is studied using UV-Vis-NIR spectrophotometer and Atomic Force Microscopy. Moreover, the ZnO nanoparticles are used as the charge trapping layer in a MOS memory device. The memory performance is studied using high-frequency C-V measurements.

Experimental

First of all, a quartz wafer is dipped into a solution containing 1-5 nm ZnO nanoparticles which are stabilized by polyacrylate sodium; then the sample is left for 2 hours to dry at room temperature. Amplitude Modulation-Atomic force microscopy (AM-AFM) measurements are conducted on the sample and the $1\times1\mu m$ height map depicted in Fig. 1 shows that the majority of the ZnO nanoparticles have an average height of 1.8 nm and a width > 15-nm. This means that there is a single layer of NPs which are agglomerated in the horizontal direction only. Moreover, it has been reported that some nanoparticles agglomerations cause a change in the properties of the nanomaterials [12-21]. In this case,

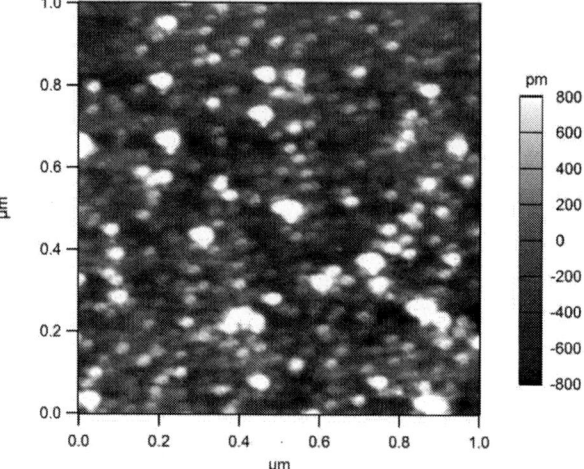

Figure 1. AFM height map ($1\times1\mu$m) of the ZnO nanoparticles agglomerations.

since the thickness of the agglomerations is in the range of the Bohr radius of the ZnO exciton (~2.3 nm) [22], 1-D quantum confinement effects are expected to occur in the agglomerations. The reflectance and transmittance spectra are measured using a UV-Vis-NIR spectroscopy. Then, to extract the bandgap of the ZnO agglomerations, the Kubelka-Munk function for a direct bandgap material is used where $(h\upsilon\alpha)^{1/2}$ vs. $h\upsilon$ is plotted as shown in Fig. 2 where $h\upsilon$ is the photon energy and α is the absorption coefficient. The extracted value is 3.7 eV which is larger than the bulk ZnO bandgap (3.4 eV). The result is consistent with the 1-D quantum confinement effects. Another expected change in the electronic properties of the ZnO due to quantum confinement in 1-D is the reduction of the electron affinity [23].

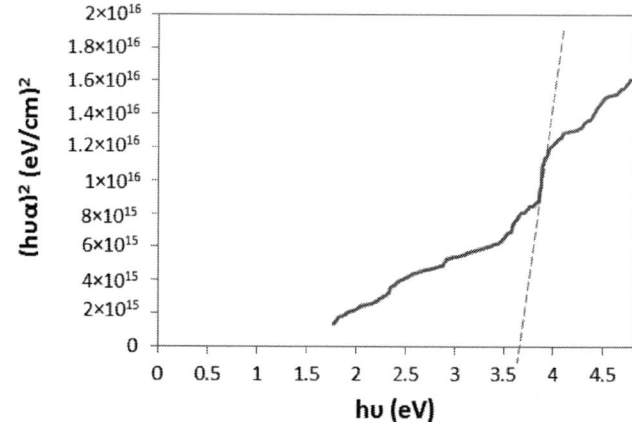

Figure 2: Schematic cross-section of the fabricated charge trapping memory cell with graphene nanoplatelets.

Then, MOS-memory devices are fabricated on an n+-type (111) (Antimony doped, 15-20 mΩ-cm) Si wafer. First, 4-nm tunnel oxide Al_2O_3 is deposited at 300°C in an Oxford FlexAL system. Then, the ZnO NPs are dip-coated on the sample which was then left to dry for 2 hours at room temperature. Next, 9-nm blocking oxide Al_2O_3 is deposited by plasma assisted ALD at 200°C. Finally, a 350-nm Al layer for the gate contact is e-beam evaporated using a shadow mask with feature size down to 10 μm which eliminated the need for any lithography steps. A cross-sectional illustration of the fabricated MOS-memory device is shown in Fig. 3.

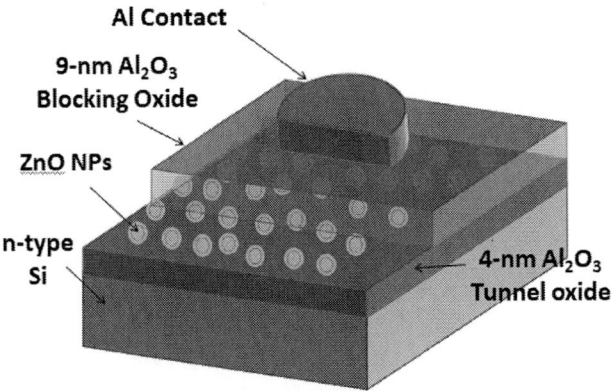

Figure 3. Schematic cross-section of the fabricated charge trapping memory cell with ZnO nanoparticles.

The electrical characterization of the fabricated memory devices is conducted by measuring the high-frequency C-V_{gate} characteristics of the programmed and erased states at 1 MHz using the Agilent-B1505A Semiconductor Device Parameter Analyzer. The memory cells gate voltage (V_g) was first swept from -10 V forward to 10 V then backwards. An 8.2 V threshold voltage (V_t) shift is measured as shown in Fig. 4.

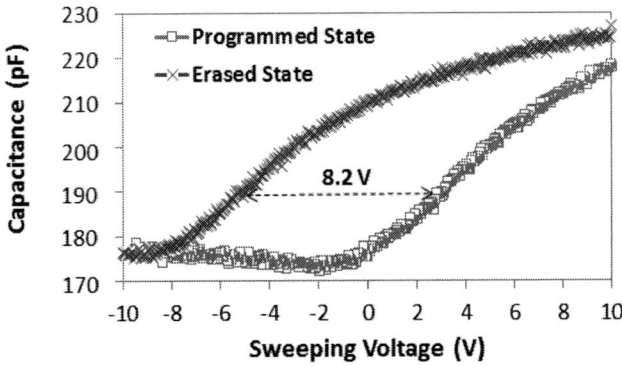

Figure 4. Schematic cross-section of the fabricated charge trapping memory cell with ZnO nanoparticles.

The C-V measurements are repeated at different gate sweeping voltages as shown in Fig. 5. The curve shows that the ZnO-NPs are providing high density charge trapping states in the 2-D quantum well, and a large memory window of 4 V is obtained at a low program/erase voltage of 7/-7 V. Moreover, at higher programming voltages, the C-V curve is found to be shifting to the right which means that holes are being trapped within the nanoislands while the erased state is almost fixed and not shifting to the right at higher erase voltages which means that negligible electrons are stored in the nanoislands. In fact, due to quantum confinement, the electron affinity of the ZnO is expected to be reduced which may inhibit the storage of electrons. However, as a result, the valence band offset between the ZnO and surrounding oxides will become larger and the quantum well will become deeper which will allow for more available energy states with high charge trapping density for the holes to be stored in.

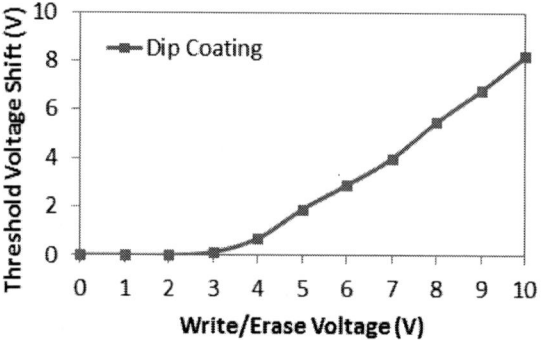

Figure 5. Measured threshold voltage of the memory with ZnO nanoparticles vs. write/erase voltage.

The charge emission-mechanism during the program-operation is also studied. The electric field across the tunnel oxide (E_{ox}) is calculated using Gauss's equation. The natural-logarithm of the V_t shift divided by the square of E_{ox} is plotted vs. the reciprocal of E_{ox} as depicted in Fig. 6. The linear trend indicates that the prevailing charge emission-mechanism at a $V_g > 5$ V (corresponding to $E_{ox} = 5.01$ MV/cm) is Fowler-Nordheim

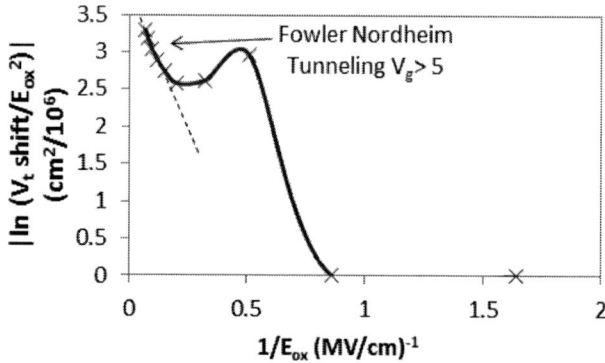

Figure 6. Measured threshold voltage of the memory with ZnO nanoparticles vs. write/erase voltage.

tunneling (F-N). In F-N tunneling, the charges are emitted by first tunneling into the conduction/valence band of the oxide through a triangular energy barrier and then are swept into the charge trapping layer by the electric field [24-28]. Also, the V_t shift is plotted vs. the square of E_{ox} and the linear trend observed in Fig. 7 indicates that Phonon-Assisted Tunneling is the dominant charge emission-mechanism at $V_g < 5$ V.

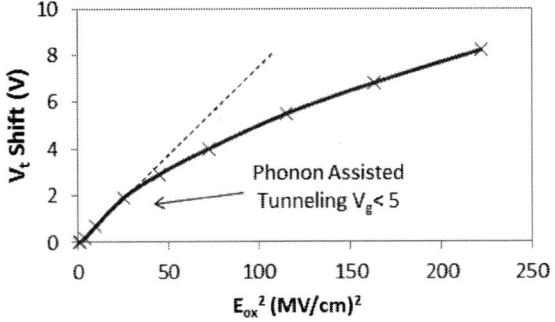

Figure 7. Measured threshold voltage of the memory with ZnO nanoparticles vs. write/erase voltage.

Furthermore, the endurance of the memory cells is characterized by programming/erasing the memory at 7/-7 V up to 10^4 cycles and observing the change in V_t shift as shown in Fig. 8. The figure shows that 18% of the initial charge is lost after 10^4 cycles where a memory hysteresis of ~3 V is shown. This good endurance characteristic is due to the good quality of the tunnel oxide [29].

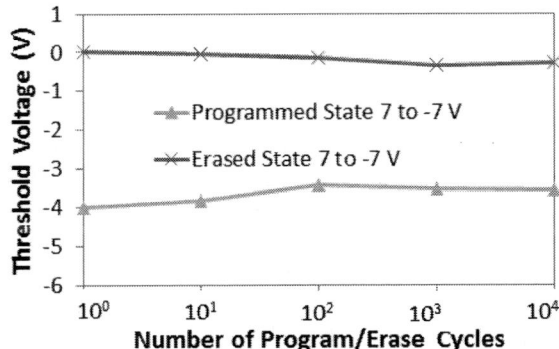

Figure 8: Endurance characteristic of the MOS memory with ZnO nanoparticles

Conclusion

In conclusion, a MOS memory device with ZnO nanoparticles deposited by dip-coating is demonstrated. The dip coating technique is shown to result in agglomerations of the ZnO nanoparticles and their bandgap is increased to 3.7 eV due to 1-D quantum confinement effect. Moreover, the memory showed holes storage in the charge trapping layer which is due to the reduction of the electron affinity of the ZnO. At low operating voltages, the holes emission mechanism is found to be based on Phonon-Assisted

Tunneling, while Fowler-Nordheim Tunneling becomes the dominant mechanism at gate voltages larger than 5 V. The results show that a large memory window can be obtained due to the large available charge trap states in the ZnO, in addition to an excellent endurance characteristic which make the ZnO nanoparticles promising in memory applications.

Acknowledgments

This work was supported by Masdar Institute of Science and Technology and the Office of Naval Research global grant N62909-16-1-2031. N. El-Atab gratefully acknowledges financial support provided by *L'Oréal-UNESCO For Women in Science* Middle East Fellowship.

References

1. U. Özgür, et al., *J. Appl. Phys.* **98**, 041301 (2005),
2. N. El-Atab, et al., *Appl. Phys. Lett.* **104**, 013112 (2014).
3. U. Koch, A. Fojtik, H. Weller, a. Henglein, *Chem. Phys. Lett.* **122**, 507 (1985).
4. N. El-Atab et al., *Appl. Phys. Lett.* **104**, 253106 (2014).
5. K. F. Lin, et al. Applied physics letters **88**, 263117 (2006).
6. D. Banerjee, J. Y. Lao, D. Z. Wang, J. Y. Huang, Z. F. Ren, D. Steeves, B. Kimball, M. Sennett, *Appl. Phys. Lett.* **83**, 2061–2063 (2003).
7. W. S. Chiua, et al., *Chem. Eng. J.* **158**, 345–352 (2010).
8. V. Polshettiwar, B. Baruwati, R. S. Varma, *ACS Nano* **3** 728–736, 2009.
9. N. El-Atab et al., *Appl. Phys. Lett.* **105**, 033102 (2014).
10. T. K. J. Gupta, et al., *J. Mater. Res.* **10**, 2295 (1995).
11. K. L. Chopra, S. Major, D. K. Pandya, *Thin Solid Films* **102**, 1 (1983).
12. P. Francis et al., *The European Physical Journal D*, **67**(7), 1-7 (2013).
13. A. R. Studart, E. Amstad, and L. J. Gauckler. Langmuir **23**.3, 1081-1090 (2007).
14. H. J. Kim, I. C. Bang, and J. Onoe, *Optics and Lasers in Engineering* **47**, 532–538 (2009).
15. Yu, Wei, and Huaqing Xie. *Journal of Nanomaterials* **1** (2012).
16. M. Rahman, et al. Berlin Heidelberg: Springer, 2013.
17. Z. L. Wang, and F. Xiangdong. *The Journal of Physical Chemistry B* **107**, 13563-13566 (2003).
18. S. Botti, et al. *Physical Review B* **78**.3 (2008): 035333.
19. F. Priya, et al. *The European Physical Journal D* **67**, 1-7 (2013).
20. F. M. Omar, H. A. Aziz, and S. Stoll. *Journal of Colloid Science and Biotechnology* **3**, 75-84 (2014).
21. L. Xuegeng, Y. He, and M. T. Swihart. *Langmuir* **20**, 4720-4727 (2004).
22. Y. Gu et al., *Appl. Phys. Lett.*, **85**, 3833 (2004).
23. H. Yu, et al. *Nature materials* **2**, 517-520 (2003).
24. N. El-Atab, et al., *Physica Status Solidi (a)* **212**, 1751-1755 (2015).
25. A. Nayfeh, et al., *ECS Trans.* **64**, 45-49 (2014) doi: 10.1149/06417.0045ecst

26. A. Nayfeh, et al., *The Electrochemical Society Meeting Abstracts*, no. 37, pp. 1879-1879 (2014).
27. N. El-Atab, *Nanoscale research letters* **10**.1, 1-7 (2015).
28. N. El-Atab, et al., 14th International Conference on IEEE Nanotechnology (IEEE-NANO), 505-509 (2014).
29. N. El-Atab, et al., 15th International Conference on IEEE Nanotechnology (IEEE-NANO), 766-768 (2015).

Chapter 5

Poster Session

82

Thermal-structural Optimization of Light with LED Packaging

L. Q. Zhang[a], L.Y. Li[b], D. H. Ge[c], X.D. Zhu[d]

[a] Laboratory of Advanced Manufacturing & Reliability for MEMS/NEMS, Department of Mechanical and Electronic, School of Mechanical Engineering, Jiangsu University, Zhenjiang, Jiangsu, 212013, PR. China
[b] FOXCONN, Baoan, Shenzhen, 518109, PR. China
[c] Research Center of Micro/nano Science & Technology, Department of Mechanical and Electronic, School of Mechanical Engineering, Jiangsu University, Zhenjiang, Jiangsu, 212013, PR. China
[d] Nanjing Taijie Electronic Technology Co., LTD., Nanjing, Jiangsu, 210000, PR. China

A systematic numerical and experimental method is developed to optimize a caution lamp with multichips LED chips. Here, we intend to describe the thermal performance of LED caution lamp with multiple LED chips through the structural optimization by the finite element method. One type of caution lamp was made and the highest temperatures of LED chips have been test after the optimization. This optimization method is useful for design and manufacturing of caution lamps with multiple LED chips. Compared with the simulation and the experiment data, the high temperature of LEDs were reduced by choosing fitting parameters. It indicates that the results and methods used in this paper can provide the guidance in the understanding of thermal management for LED lamps design.

Introduction

At present, the light-emitting-diode (LED) has been a tremendous surge in its application as a new type of generation lighting source. The thermal performance and structural reliability of LED system have been investigated by different conditions in electronic package (1-3). With the minimization of the LED packaging, the reliability problem, especially, the thermal reliability becomes more and more seriously. Hence the heat dissipation for LED system becomes much crucial (4-6). Lots of heat management methods have been investigated for LED system designing and manufacturing. The most popular cooling method used in the LED system is the passive cooling, however, its cooling efficiency is very low. It is very significant to develop the heat dissipation method (7-10). The paper is intended to describe the thermal performance of LED caution lamp with multiple LED chips through the structural optimization. This optimization method is useful for design and manufacturing of caution lamps with multiple LED chips.

Models and Simulation

Physical Models

Figure 1 shows the internal structure of caution lamp, it contains the LED chips, lamp post, and heat dissipation base and lamp socket. In the simulation, these three structural parameters, r (the radius of lamp post, which cross section is hexagon), D (the diameter of the heat dissipation base) and (h) the height of the heat dissipation base, are chosen. At the beginning, the whole height of this part is 132mm. The finite element model of this caution lamp is built based on this physical model.

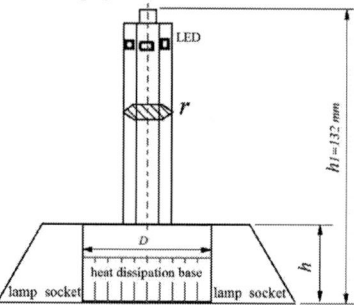

Figure 1. Physical structural model of caution lamp with LEDs

FE Model and Initial Boundary Conditions

Before the simulation, the initial conditions used in this simulation is as follows, the references environment temperature is set as home temperature 25 ℃; The natural convection coefficient applied on the lens surface is about 2 $W/m\cdot K$; the thermal power of each LED is 2 W; the initial value of each parameter are as follows, r =4 mm; D=100 mm; h=40 mm. In order to understand the relationship between the highest temperature on LED or caution lamp and the structural parameters, a simple finite element model is built. The thermal conductivity of each materials used in this model is as follows, the thermal conductivity of GaN-LED is equal to 17 $W/m\cdot K$; the thermal conductivities of Lamp post and heat dissipation base (aluminum 1100) are equal to 218 $W/m\cdot K$; the thermal conductivities of Substrate (copper clad aluminum plate) is equal to 154 $W/m\cdot K$. The parameter optimization combination must be obtained to meet the requirement of the design temperature does not exceed 70℃. It can be described by one function as follows,

$$T_{\max}(r, h_1, D, h) \leq 70℃$$

$$s.t.$$

$$\begin{cases} 0 < r \leq 10 \\ 0 < h \leq h_1 \\ r \leq D \leq 160 \\ h_1 = 132 \end{cases} \qquad [1]$$

Numerical Simulation Results

According to the results, the final parameters of this lamp is as follows, r=6mm; D=150mm; h=43mm; h_1=132mm. The maximum temperature of this lamp is about 69.67 ℃, while the maximum temperature of this lamp is about 89.23 ℃ before optimization.

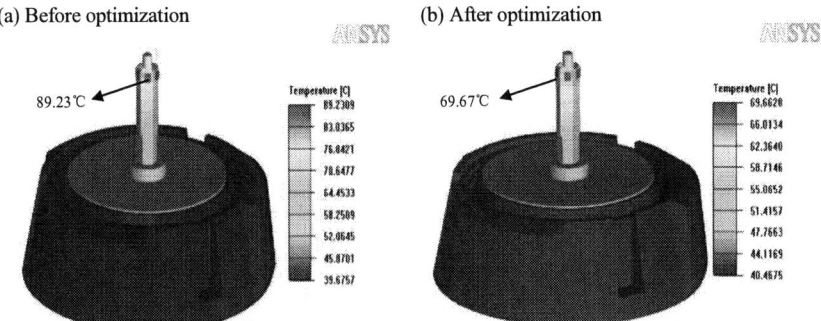

Figure 2. The temperature distribution of caution lamp before and after optimization

Experiment Results

According to the simulated results, the caution lamp sample with six LED chips has been prepared by the optimization parameters. The thermalcouples are used to test the temperature of the caution lamp. Figure 3 (a) shows the connection principle based on the Thermalcouples. In the experiment, the Thermalcouples are adhesive on the LED, then; the temperature can be obtained by using the data collector. The experimental test pattern is shown in Figure 3 (b). In the test, we did four test of each point (I, II, III) in the case of electricity 24V.

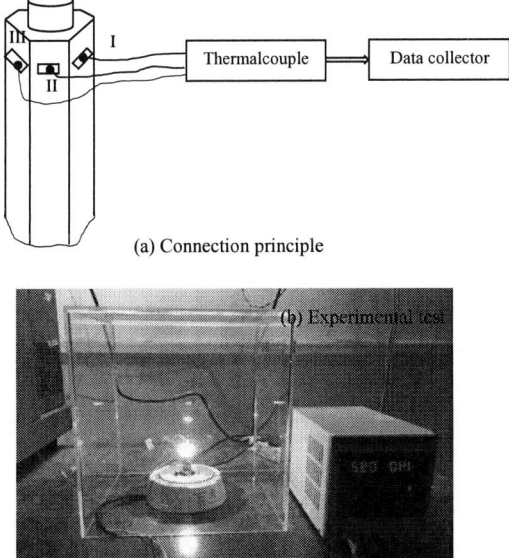

Figure 3. The temperature measurement principle and measuring system

Discussions and Conclusions

According to the simulation and the experimental date, The highest value is 66. 84 ℃ of the II point at the fourth time test shown in Figure 4. However, the lowest value is 63. 83 ℃ of the III point at the first time test. The error of the test is about 1.51%, which implies that the change trend is not obvious. It reveals that the lamp system is in the thermal balance situation. According to the simulation results, the average temperature of LEDs is 69.66 ℃, while, the average result of the test is 65.25, therefore, the error between the numerical simulation and the test is about 6.76%.

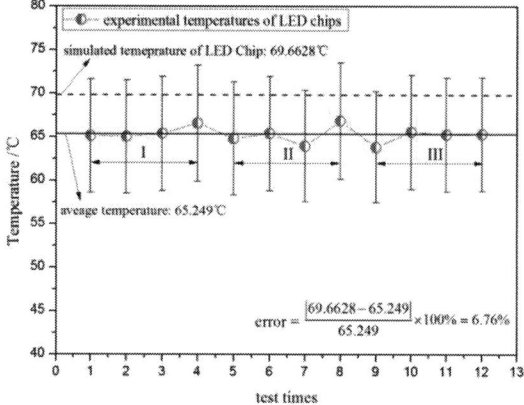

Figure 4. experiemenatal and simulated resutls

The understanding of the thermal characteristics of the LED packaging or assemblies is very important for designing efficient and high-power LEDs. A systematic numerical and experimental method is developed to optimize a caution lamp with multichips LEDs in this paper. According to the comparison on the simulation and test, we can draw some conclusions as follows, in the structural design of caution lamp with LED chips, the parameter of heat dissipation base diameter is the most important factor compared with other parameters; The highest temperature on LED chips can be reduced through increasing the structural size, which can also improve the reliability of lamps system; The systematic method used in this paper is benefit for the guidance to understand the thermal management for LED application.

Acknowledgments

The authors would like to acknowledge the support of natural science foundation of Jiangsu province youth fund (BK20130537 & No. BK20140556), the national natural science foundation of China (No.11404146), the China postdoctoral science foundation (No.2014M561575), the senior talent start-up foundation of Jiangsu university (13JDG020 & 13JDG021), the support of Jiangsu province postdoctoral foundation (1402168C & No.1301047C), and the specialized research fund for the doctoral program of higher education of China (No. 20133227120022) during the course of this work. The authors also would like to acknowledge Nanjing Taijie electronic technology Co. LTD. for her manufacturing and supplying of these samples.

References

1. D.J. Liu, H.Y. Yang, P. Yang, *Microelectron. Reliab.,* **54**, 926(2014).
2. M. Meneghini, L.R. Trevisanello, G. Meneghesso, E.Z. Zanoni, *IEEE Device Mat. Re.*, **8**, 323(2008).
3. M.Meneghini, T.Augusto, M. Ciovanna, M. Gaudenzio, Z. Enrico, *IEEE Trans. Electron Dev.,* **57**, 108 (2010).
4. S.F. Sufian, Z.M. Fairuz, M. Zubair, M.Z. Abdullah, J.J. Mohaned, *Microelectron. Reliab.,* **54**, 1534(2014).
5. M.H. Chang, D. Das, P.V.Varde, M. Pecht, *Microelectron. Reliab.,* **52**, 762(2012).
6. J.H. Choi, M.W. Shin, *Microelectron. Reliab.,* **52**, 830(2012).
7. D. Jang, S.H. Yu, K.S. Lee, *Int. J. Heat Mass Tran.*, **55**, 515(2011).
8. B.H. Kim, C.H. Moon, *IEEE T. Compon. Pack.*, **2**, 1832(2012).
9. V.A.F. Costa, A.M.G. Lopes, *Appl. Therm. Eng.*, **70**, 131(2014).
10. S.J. Park, Y.L. Lee, *Transactions on Electrical and Electronic Materials*, **15**, 201(2014).

88

Mechanical Analysis of Stretchable AlGaN/GaN High Electron Mobility Transistors

R. P. Tompkins[a], I. Mahaboob[b], F. Shahedipour-Sandvik[b], N. Lazarus[a]

[a] Sensors and Electron Devices Directorate, U.S. Army Research Laboratory, Adelphi, Maryland 20783, USA
[b] College of Nanoscale Science and Technology, SUNY Polytechnic Institute, Albany, New York 12203, USA

Adding waves or kinks to a conductor is a well-known method for relieving strain in stretchable systems. Although transistors and other devices made of brittle semiconductors have also been demonstrated, the effects of the crystalline anisotropy on the mechanical behavior has not been previously investigated. In this work, a numerical solver is used to simulate the peak stresses in silicon and gallium nitride embedded within soft silicone in common stretchable geometries. As expected from other work, adding waves reduces peak stress by more than a factor of ten. The effects of in-plane rotations were then investigated. In the typical [0001] growth direction for gallium nitride, the matrix is rotationally invariant, with no variation in peak stress upon rotation about the [0001] direction. For silicon, however, peak stress was found to vary by as much as 17 percent for rotation about the [001] direction.

Introduction

Wearable electronics systems intended to be mounted on human skin must survive tens to hundreds of percent strain without fracture [1]. However, semiconductor materials such as Si, GaAs, and GaN used for electronic devices are rigid in nature and are not intrinsically stretchable. The high temperature required to grow these materials as well as the amorphous/polycrystalline nature of stretchable substrates such as silicone and polyurethane also does not render direct growth of device layers on substrates a viable approach. One possibility is to grow sufficiently thin device layers (1 - 5 μm) on rigid substrates, undercut the device through a selective etch, embed the device in a stretchable material followed by removal of the substrate or wafer thinning process [1]. Stretchable devices fabricated using this approach must also exist in unique geometries to accommodate the large amount of additional strain, where conventional straight devices would lead to fracture due to stress concentration. In addition to looking at silicon devices, we are also interested in investigating the possibility of instead using GaN transistors for stretchable power devices. Resistive losses in GaN devices are much less than that of Si due to the wider band gap and thus higher critical field [2]. Resistive losses and resulting heat generation are of particular importance for wearable systems where an increase in temperature of < 5 °C can cause discomfort to the user [3]. Thus, use of GaN-based electronic devices for stretchable electronics is of particular interest.

In this work we perform mechanical simulations using the finite element solver COMSOL Multiphysics 3.3 to examine the peak von Mises stress in GaN or Si with stretchable device geometries after application of 30% strain parallel to the trace. The effect of crystal orientation in both GaN and Si are then investigated for the first time by simulating with rotations of the crystalline plane relative to the stretchable geometry in each case.

Experimental

Stress in a material is governed by Hooke's law for continuous media:

$$\sigma = C\varepsilon \qquad [1]$$

where σ is the stress, C is the stiffness tensor and ε is the applied strain. The values of the stiffness matrix used in this work are found in reference 4 for GaN and reference 5 for Si. In each calculation, we applied a 30% mechanical strain parallel to the trace and subsequently recorded the peak von Mises stress. The mesh size was between 11,320 and 22,401 mesh points depending on geometry. Mesh points for one period are shown in Figure 1 for a 1 mm peak-to-peak amplitude sinusoid. Mesh points are more concentrated around areas that have the largest change in curvature where the peak stress is most likely to be found.

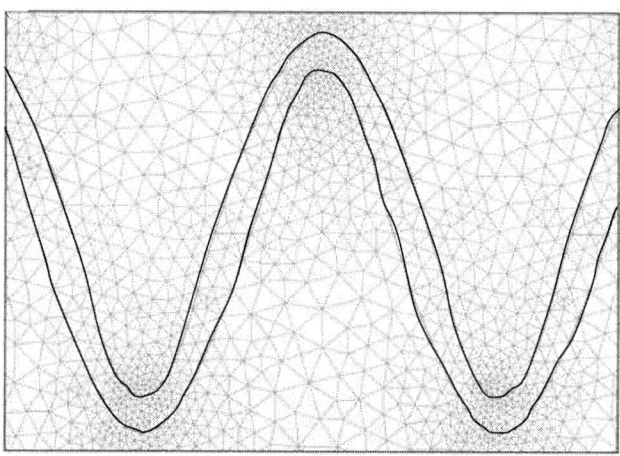

Figure 1. Image showing the mesh used in finite element analysis. Geometry shown is the sinusoidal geometry with 1 mm peak-to-peak amplitude. The trace is GaN or Si material. Material surrounding the trace is Sylgard 184.

The 2-D structure we simulated consists of GaN or Si semiconductor traces surrounded by Sylgard 184 silicone commonly used in stretchable electronics in a rectangular structure of dimensions 3 mm x 10 mm. A 30% strain was applied parallel to the trace on one end of the rectangle. The opposite end of the applied strain parallel to the trace was fixed. The two long sides of the rectangular structure were permitted to move freely. Thicknesses for both the GaN and silicone were 5 microns. Based on previous simulations for copper found in reference 6, we looked at the following stretchable geometries: straight, curved-corner rectangular, horseshoe, rectangular, and sinusoidal shown in Figure 2. Results were compared to the conventional straight geometry. Geometries for each shape included a total of 10 periods, with varying peak-to-peak amplitude of 0.25, 0.5, and 1.0 mm. For comparison, all simulations were then repeated with Si as the semiconductor trace material.

Figure 2. Trace geometries used in this work, from left to right: straight, curved-corner rectangular, horseshoe, rectangular, and sinusoidal.

Results

<u>(0001) GaN and (100) Si</u>

Figures 3 and 4 shows an example for the output of the COMSOL Multiphysics program for GaN material with a with peak-to-peak amplitude of 1 mm for both sinusoidal and rectangular geometries. The plot shows that regions with the highest stress are concentrated where the curvature changes the most, near the maxima and minima of the trace for the sinusoid geometry and regions parallel to the trace for the rectangular geometry. The peak von Mises stress was 12 GPa after 30% applied strain parallel to the trace for the sinusoid configuration and 7.7 GPa for the rectangular configuration.

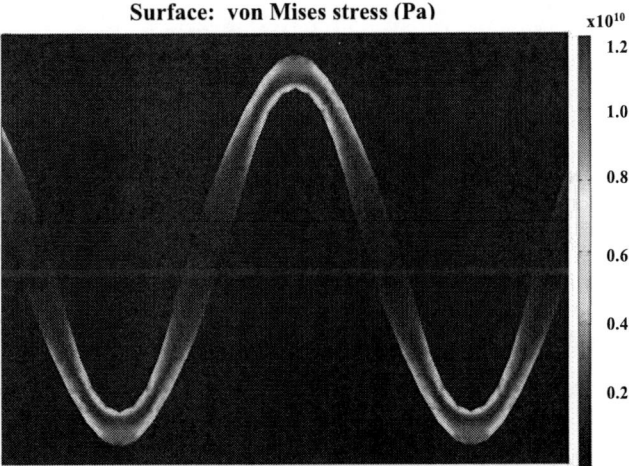

Figure 3. Output from the COMSOL showing von Mises stress for sinusoidal geometry peak-to-peak amplitude of 1 mm.

Surface: von Mises stress (Pa)

x10^9

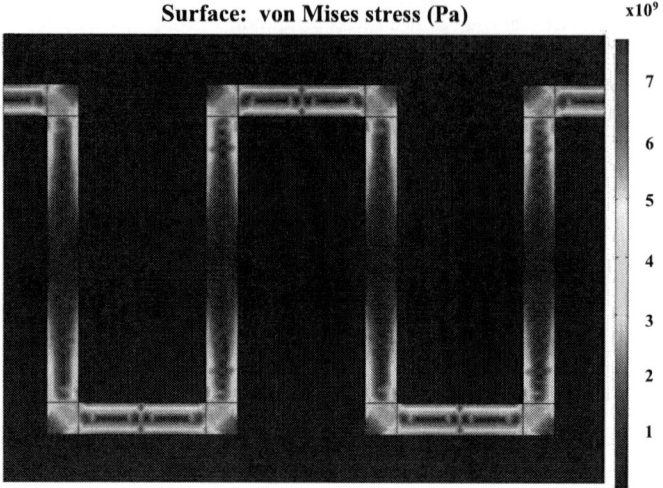

Min: 4.746 x 10^4

Figure 4. Output from the COMSOL showing von Mises stress for sinusoidal geometry peak-to-peak amplitude of 1 mm.

Figure 5 is a plot of all simulations in multiple geometries for (0001)GaN as well as (100) Si. Inspection of the results in Figure 5 show a number of trends and observations. First, as expected, for each geometry irrespective of material, there is a decrease in peak stress as the arc length increases. In most cases, the peak stress found in Si is less than that in GaN, most likely associated with the fact that GaN is more rigid than Si, with values for the stiffness matrix along the principal axes, (C_{11}, C_{22}, and C_{33}) approximately twice as large for (0001)GaN [4] compared to (100)Si [5]. In this work, we limited our calculations to geometries that were the most promising for stretchable inductors found in reference 6. Results shown in Figure 5 show 30% − 98% reduction in peak stress in GaN by going to stretchable geometries compared to the conventional straight geometry, while a 28% − 92% reduction is observed in Si. It is most likely all the designs in this study are potential approaches for stretchable electronics, however, our results could vary with further discretization of the individual curves, in particular for traces that have a gradual change in curvature such as the sinusoidal case. It is possible that such a calculation will give slightly different results than what we computed here and thus one specific geometry may be preferred over all others. However, discretization is also an issue in mask writing for contact photolithography and thus rounded features are never perfectly round. Further investigation will look at discretization on a finer scale on the order of mask resolution.

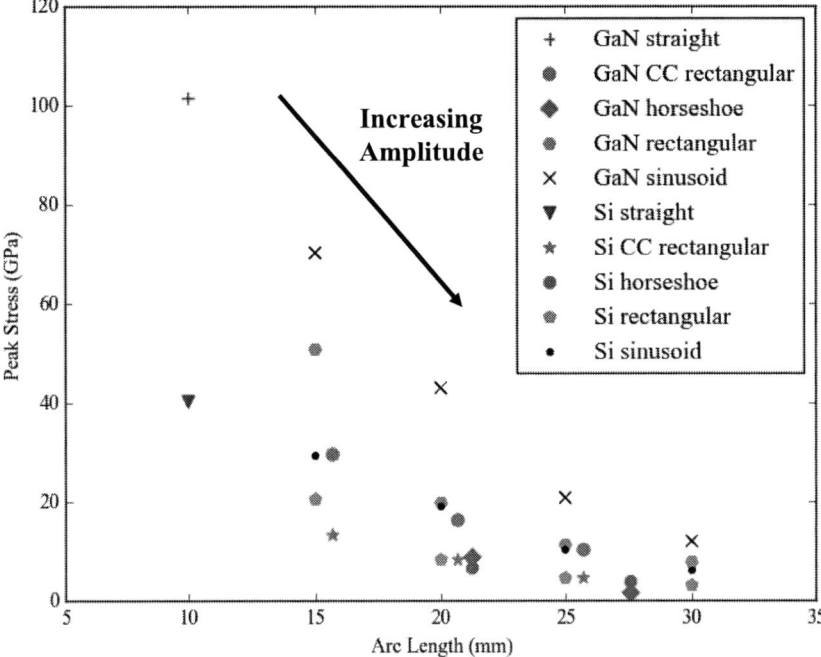

Figure 5. Plot of peak stress vs. arc length for multiple trace geometries for both (0001)GaN and (100)Si material [7]. The arrow indicates direction of increasing peak-to-peak amplitude.

In-plane rotational dependence

(0001)GaN is hexagonal in nature and thus the stiffness tensor (C) is rotationally invariant with in-plane rotation (rotation about the [0001] direction). This property of (0001)GaN is one advantage over (100)Si for stretchable electronics, where Si is cubic and thus does not have the same rotational symmetry. We examined Si with rotations in 10 degree increments from 0 to 90 degrees. We also included the high symmetry 45 degree rotation. We limited our study to two specific geometries, the sinusoidal geometry and rectangular geometry both with a 1 mm peak-to-peak amplitude where a 30% mechanical strain was applied parallel to the trace and the peak von Mises stress was recorded. Results from our calculations are shown in Figure 6. Results show the expected constant values for the peak stress in GaN for both geometries. Variation in the Si was 12% for the sinusoid geometry and 17% for the rectangular geometry. We observe different trends in peak stress vs. rotation angle for the two geometries. The location of the peak stress in the rectangular geometry is parallel to direction of applied strain (shown in Figure 4) and hence the maximum peak stress is at the 45 degree rotation when the stiffest axis is aligned along this direction. For the sinusoidal trace, on the other hand, the stress peaks along an angled portion of the curve (Figure 3), resulting in a stress minimum at a roughly 45 to 55 degree rotation to the direction of applied strain.

Because we are most interested in HEMT devices for stretchable applications, we limit our simulations in this investigation to (0001) GaN where c-axis growth of GaN is a requirement for maximum polarization charge and thus high two-dimensional electron gas charge concentration. However, both non-polar and semi-polar GaN orientations are common for LEDs to reduce or eliminate the effects of polarization fields on carrier recombination [8]. Future calculations will examine these orientations.

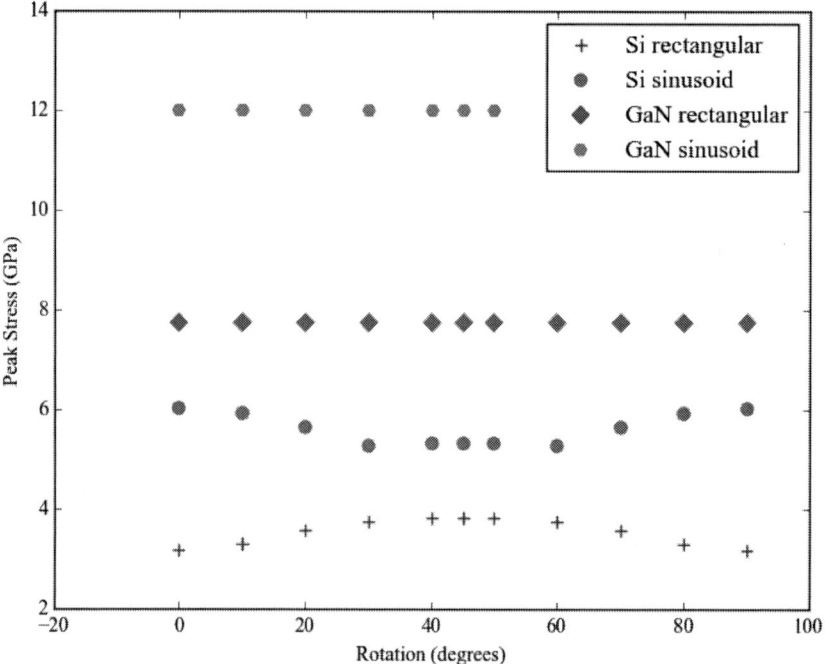

Figure 6. Plot of peak stress vs. rotation angle. All geometries have a peak-to-peak amplitude of 1 mm [7].

Conclusion

In this work, we used COMSOL Multiphysics 3.3 to compute the peak stress in both GaN and Si with multiple trace geometries embedded in Sylgard 184 as a stretchable substrate. We limited our study to trace shapes that are possible candidates to fabricate stretchable devices. Our calculations suggest that all geometries are potential designs for stretchable electronics as our calculations suggest a 30% – 98% reduction in peak stress in GaN by going to stretchable geometries compared to the conventional straight geometry, while a 28% – 92% reduction is observed in Si. However, these results should be approached with caution as further discretization could render different values of the peak stress in particular for geometries that have a gradual change in curvature. Our results also confirm the expected results that Si had a lower peak stress than GaN as well as a decreasing peak stress with increasing peak-to-peak amplitude. We then examined the effects of in-plane rotation on the peak stress a function of angle for both materials. One advantage of (0001) GaN over (100) Si is that the stiffness matrix is rotationally invariant in-plane. This property of GaN is of particular importance for contact lithography where misalignment has no effect on the stiffness tensor. Variation in Si in rotating from 0 – 90 degrees in 10 degree increments was approximately 17% for the rectangular trace and 12% for the sinusoidal geometry, both having a 1 mm peak-to-peak amplitude.

References

1. D. Kim, N. Lu, R. Ma, Y. Kim, R. Kim, S. Wang, J. Wu, S. M. Won, H. Tao, A. Islam, K. J. Yu, T. Kim, R. Chowdhury, M. Ying, L. Xu, M. Li, H. Chung, H. Keum, M. McCormick, P. Liu, Y. Zhang, F. G. Omenetto, Y. Huang, T. Coleman and J. A. Rogers, *Science*, vol. 333, 2011, 838.
2. L. Bin, D. Piedra, and T. Palacios, *8th International Conference on Advanced Semiconductor Devices and Microsystems*, (2010)105-110.
3. H. Fruhstorfer, U. Lindblom, and W.G. Schmidt, J. Neurology, *Neurosurgery, and Psychiatry*, 39, (1976) 1071-1075.
4. M. Rais-Zadeh, Mina et al., *Journal of Microelectromechanical Systems* 23.6 (2014): 1252-1271.
5. J. Singh, "Electronic and Optoelectronic Properties of Semiconductor Structures," Cambridge: Cambridge University Press pg. 35 (2003).
6. N. Lazarus, C. D. Meyer and S. S. Bedair, *IEEE Trans. on Electron Devices*, **62** 2270 (2015).
7. J.D. Hunter, *Comput. Sci. & Eng*, **9** 90 (2007).
8. H. Masui, S. Nakamura, S. P. DenBaars and U. K. Mishra, *IEEE Transactions on Electron Devices*, vol. 57, no. 1, pp. 88-100, Jan. 2010.

96

ECS Transactions, 72 (5) 97-108 (2016)
10.1149/07205.0097ecst ©The Electrochemical Society

Effect of Ba co-doping on $Sr_5(PO_4)_3Cl:Ce^{3+}$ Blue Emitting Phosphor

Application for White Light Emitting Diodes

G. Deressa

Department of Chemistry, School of Natural Science, Adama Science and Technology
University, Adama, 1888, Ethiopia

Ce^{3+}-activated $Sr_{5-2x-y}Ba_y(PO_4)_3Cl:xCe^{3+},xNa^+$ phosphors were synthesized via the solid statereaction methods. The substitution of Ba^{2+} for Sr^{2+} ions was confirmed by Raman spectra. The broadest Raman width in $Sr_{2.45}Ba_{2.45}(PO_4)_3Cl:Ce^{3+}$ phosphor supported its highest disorder of the host lattice, leading to the emission spectral broadening with a half width of 95.5 nm due to the larger variation in crystal filed strength. The thermal stability of the luminescence was evaluated by the temperature-dependent luminescence intensity. The simulating a spectra of blue-emitting $Sr_{2.45}Ba_{2.45}(PO4)_3Cl:0.05Ce^{3+},0.05Na^+$ and orange emitting $Sr_3SiO_5:Eu^{2+}$ phosphors as light converters, an intense white GaN-based near-UV-LED (400 nm) was exhibit good color-rendering index Ra of 90.8 at a correlated color temperature of 6473 K and CIE coordinates of (0.3165, 0.3051). These characteristics makes $Sr_{2.45}Ba_{2.45}(PO_4)_3Cl: 0.05Ce^{3+},0.05Na^+$ should be a promising blue-emitting phosphors for the production of white light in phosphor conversion white-LEDs.

Keywords: Blue apatite phosphor; Spectral broadening; white-LED

---

*Corresponding Author's E-mail; adamachemistry@gmail.com, Mobil; +251968062510

1. Introduction

Solid-state lighting has received intense and growing interest over the past few years, due to light emitting diodes (LEDs) properties of high luminous efficiency, long lifetimes, and low pollution and power consumption. The commercialized white light emitting diodes (white-LEDs) can be generated by combining a blue-emitting InGaN LED chip and a yellow-emitting garnet phosphor $Y_3Al_5O_{12}:Ce^{3+}$ (YAG:Ce^{3+}) [1]. However, such white light does not have sufficient color rendering properties. The color-rendering index (CRI) remains to be improved due to the lack of a full color component from the yellow phosphor. Thus, tricolor (blue, green and red) phosphor-based white-LEDs have been suggested as a solution [2].

The phosphors activated by rare-earth ions like; Ce^{3+} and Eu^{2+}, attract the attention of researchers due to the parity-allowed transitions between the 4f and 5d levels. The 5d state is an outer orbital; the energy associated with an electron in the 5d state depends strongly on the nature of the crystal field. Hence, the lanthanide 5d energy will vary remarkably. The absorption wavelengths of 4f–5d transitions in Ce^{3+} range from UV to blue regions depending on a crystal-field splitting of the 5d levels, which is determined by the local structure of a host crystal [3,4].

Changing the cation ratio of apatite-phosphate can distorted the lattices structure due to change in crystal field of the host cases the broadening spectra of the luminescence are a promising hosts for lighting and display due to its good thermal stability and high luminescent efficiencies. The Ce^{3+}-activated apatite-phosphates are good candidates as host structures and offer a number of merits, such as high chemical and low synthesis temperature and physical stability, and they exhibit interesting luminescence. Examples include $Sr_5(PO_4)_2(SiO_4):Ce^{3+}$, $Ba_5(PO_4)_3Cl:Ce^{3+},Eu^{2+}$, $Sr_5(PO_4)_3Cl:Ce^{3+},Eu^{2+}$, $Ca_5(PO_4)_3X$: Ce^{3+} (X = Cl, F), and $M_5(PO_4)_3F:Ce^{3+}$ (M= Ba, Sr, Ca) phosphors are reported so far [5,6], there are not so many reports on Ce^{3+} activated apatites phosphors.

In the present paper, we report a broader blue emitting $Sr_{5-2x-y}Ba_y(PO_4)_3Cl:xCe^{3+}xNa^+$ phosphor. The optical properties of $Sr_{5-2x-y}Ba_y(PO_4)_3Cl:xCe^{3+}xNa^+$ were systematically investigated by means of photoluminescence excitation (PLE) and emission (PL) spectra, thermal stability and applications for near ultraviolet white light emitting diodes (near-UV-white-LEDs).

2. Experimental

Powder samples of $Sr_{5-2x-y}Ba_y(PO_4)_3Cl:xCe^{3+},xNa^+$ were prepared by a solid-state reaction at 1100 °C under a reducing atmosphere. The prepared samples were identified by X-ray diffraction (XRD) analysis using a Rigaku D/MAX 2500 with Cu Kα radiation. The measurements of photoluminescence (PL) and photoluminescence excitation (PLE) spectra were carried out using a DARSA PRO-5200 fluorescence spectrophotometer equipped with a xenon lamp as the excitation light source. The thermal quenching characteristics were measured in the temperature range of 25-200 ℃. Raman spectra of $Sr_{5-2x-y}Ba_y(PO_4)_3Cl:xCe^{3+}xNa^+$ were carried out at room temperature in the range of 200 to 2000 cm^{-1} using Agiltron, Raman Spectrometer, which is having excitation Nd:YAG laser) a excitation laser source (1064 nm).

3. Results and discussion

XRD patterns of $Sr_{5-2x-y}Ba_y(PO_4)_3Cl:xCe^{3+},xNa^+$ with (x = 0.05 and y = 0, 2.45 and 4.9 mole) are shown in Figure 1. It is clearly observed that the as synthesized sample is well coincident with of $Sr_{4.9}(PO_4)_3Cl$: $0.05Ce^{3+}0.05Na^+$(JCPDS No. 16-0666) [3], $Sr_{2.45}Ba_{2.45}(PO_4)_3Cl$: $0.05Ce^{3+}0.05Na^+$(JCPDS No. 87-1515) [3,4] and $Ba_{4.9}(PO_4)_3Cl$: $0.05Ce^{3+}0.05Na^+$(JCPDS No. 70-2318) The structure of $Sr_{5-2x-y}Ba_y(PO_4)_3Cl:xCe^{3+},xNa^+$crystals are having apatite type $M_5(PO_4)_3Cl$ (M = Ca, Sr and Ba) hosts [3,4].

Figure 2 shows room temperature normalized PL spectra of $Sr_{5-2x-y}Ba_y(PO_4)_3Cl$: xCe^{3+},xNa^+ (x = 0.05, y = 0, 2.45 and 4.90 mole) phosphors for an excitation wavelength of 365 nm. The emission band at 446 nm, 449 nm and 441 nm with full width at half maximum (FWHM) of 37.6 nm, 95.5 nm and 58.9 nm of the synthesized $Sr_{4.9}(PO_4)_3Cl:0.05Ce^{3+},0.05Na^+$, $Sr_{2.45}Ba_{2.45}(PO_4)_3Cl:0.05Ce^{3+},0.05Na^+$ and $Ba_{4.9}(PO_4)_3Cl:0.05Ce^{3+},0.05Na^+$ phosphors, respectively, are originates electron transition of Ce^{3+} ions in the hosts. [5,6].

The positions of the $4f^05d^1$ levels are much more influenced by the outer crystal field interaction than the $4f^1$ levels and highly depend on the crystalline environment around Ce^{3+} ion, so significant optical changes are expected if local structure around the Ce^{3+} center is different [4, 5]. The PL spectrum of the dopant Ce^{3+} ion experienced weak crystal fields (blue shift) and low symmetry (broaden) in $Ba_5(PO_4)_3Cl$ than $Sr_5(PO_4)_3Cl$ phosphors, due to local structure (symmetry) difference around Ce^{3+} center in the respective hosts.The lattices structure of the mixed host can be distorted due to the formation of different kinds of sites and lowering the symmetry of the hosts compared with $Sr_5(PO_4)_3Cl$ and $Ba_5(PO_4)_3Cl$ [3,4]. The more various environments of Sr-Ba(I) and Sr-Ba(II) sites in $Sr_{2.45}Ba_{2.45}(PO_4)_3Cl:0.05Ce^{3+},0.05Na^+$ phosphor means the existence of more kinds of Ce^{3+}(I) and Ce^{3+}(II) sites, which surrounded by different type of ligands. It indicates that more kinds of Ce^{3+}(I) and Ce^{3+}(II) ions experience with variations of crystal field strength. Therefore, Ce^{3+} ions in lower symmetry $Sr_{2.45}Ba_{2.45}(PO_4)_3Cl$ influences by the strong crystal fields of the neighbors surrounding Sr-Ba(I) or Sr-Ba(II) sites which can affect the electron of 5d state of Ce^{3+} ions and causes a redshift of $Sr_{2.45}Ba_{2.45}(PO_4)_3Cl:0.05Ce^{3+},0.05Na^+$ phosphor.

It can be seen from Figure 3 that the $Sr_{5-2x-y}Ba_y(PO_4)_3Cl:xCe^{3+},xNa^+$ (x = 0.05, y = 0, 2.45 and 4.90) phosphors monitored at 450 nm shows several excitation bands in the region from 250 to 420 nm, which attributed to the transition from the $4f^1$ ground state to the different crystal field splitting levels of the $4f^0 5d^1$ state for doped Ce^{3+} ions in the $Sr_{5-2x-y}Ba_y(PO_4)_3Cl$ host, indicating that the phosphor could match the near-UV LED chips (350-420 nm).

The shift in the observed Raman frequencies is believed to be due a tightening or enlargement of the molecular PO_4^{3-} ion as oxygen atoms of neighboring molecules are either drawn near or pushed apart on substitution of cations of different size Figure 4 and Table 1. Both XRD and Raman spectra of the host $Sr_{2.5}Ba_{2.5}(PO_4)_3Cl$ found between the host of $Ba_5(PO_4)_3Cl$ and $Sr_5(PO_4)_3Cl$ lattices. This indicate that the lattices of $Sr_{2.5}Ba_{2.5}(PO_4)_3Cl$ host combined from an expanded Ba-O and contracted Sr-O bonds which create different environments around the sites of Ce^{3+} ions doped in the lattice of $Sr_{2.5}Ba_{2.5}(PO_4)_3Cl$ phosphor and cause for the red shifts and spectral broadening of Ce^{3+} activated $Sr_{2.45}Ba_{2.45}(PO_4)_3Cl$ phosphors[4,7].

Conclusion

The broader blue emitting phosphor $Sr_{5-x-y}Ba_y(PO_4)_3Cl:xCe^{3+}$ (x = 0.1, and $0 \leq y \leq 4.90$ mole) phosphorshave been synthesized successfully by solid state reaction and investigated its structural and luminescence properties. Moreover, it shows that these materials can be broader under ultraviolet radiation by varying the ratio of Ba^{2+}/Sr^{2+} in the host matrix. Thus $Sr_{5-x-y}Ba_y(PO_4)_3Cl:xCe$ Phosphors can be potentially useful as a UV radiation-converting blue phosphor for fabrication of high color rendering index white-LEDs.

The white-LEDs based on near-UVchip with blue-emitting $Sr_{2.45}Ba_{2.45}(PO_4)_3Cl:0.05Ce^{3+},0.05Na^+$and orange-emitting $Sr_3SiO_5:Eu^{2+}$ phosphors were simulated with Light Tools (Version 8.3) Figure.5 [3]. A mixture of broad blue-emitting

$Sr_{2.45}Ba_{2.45}(PO_4)_3Cl:0.05Ce^{3+},0.05Na^+$ and orange-emitting $Sr_3SiO_5:Eu^{2+}$ phosphors were selected in conjunction with 400 nm chip to simulate white-LED devices. The CIE color coordinates, correlated color temperature (CCT) and average color-rendering index (Ra) of the simulated white-LEDs were found to be (x, y) = (0.3165, 0.3051), 6473K and 90.8, respectively. The results indicated that it was a promising candidate as a blue-emitting phosphor for the application of high color rendering index white light-emitting diodes.

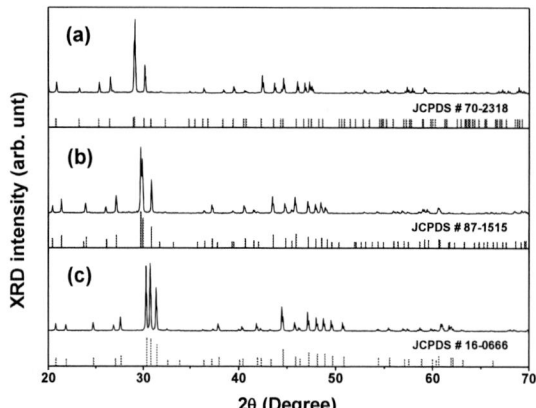

Figure 1. XRD patterns of the blue $Ba_{4.9}(PO_4)_3Cl:0.05Ce^{3+},0.05Na^+$ (a) $Sr_{2.45}Ba_{2.45}(PO_4)_3Cl:0.05Ce^{3+},0.05Na^+$(b)and$Sr_{4.9}(PO_4)_3Cl:0.05Ce^{3+},0.05Na^+$(c) phosphors.

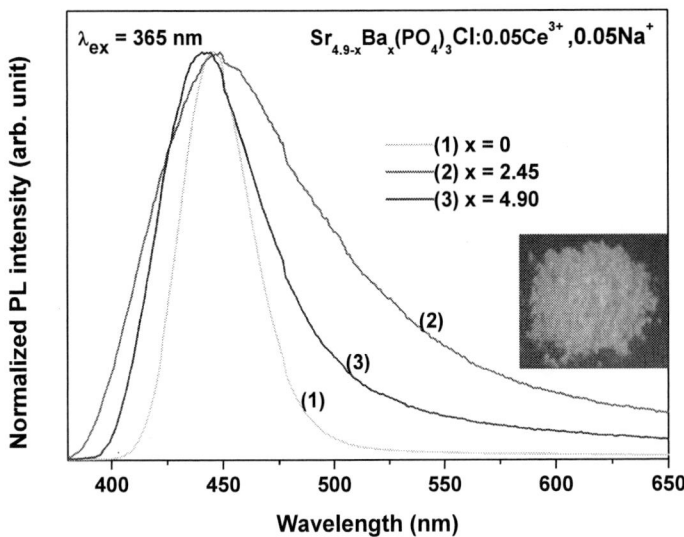

Figure 2. PL spectra of $Sr_{5-2x-y}Ba_y(PO_4)_3Cl:xCe^{3+},xNa^+$ (x = 0.05, y = 0, 2.45 and 4.90 mole) phosphors under 365 nm excitation. Insert shows the real image of the phosphors under UV.

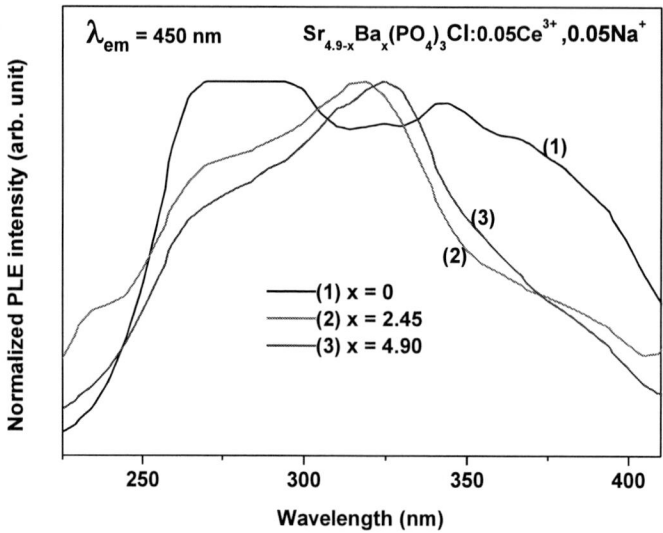

Figure 3. PLE spectra of $Sr_{5-2x-y}Ba_y(PO_4)_3Cl:xCe^{3+},xNa^+$ (x = 0.05, y = 0, 2.45 and 4.90 mole) phosphors monitored at 450 nm.

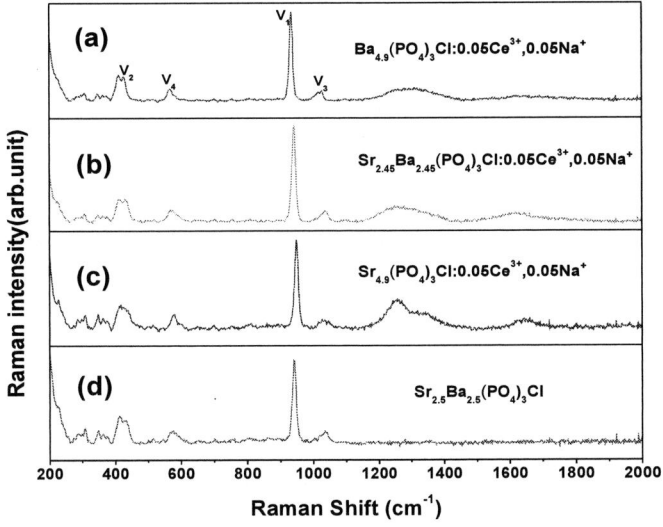

Figure 4. Raman spectra of the blue $Ba_{4.9}(PO_4)_3Cl:0.05Ce^{3+},0.05Na^+$ (a), $Sr_{2.45}Ba_{2.45}(PO_4)_3Cl:0.05Ce^{3+},0.05Na^+$ (b) $Sr_{4.9}(PO_4)_3Cl:0.05Ce^{3+},0.05Na^+$ (c) and $Sr_{2.45}Ba_{2.45}(PO_4)_3Cl$ (d) phosphors.

Table 1. Raman shifts of PO_4 modes in $Sr_{5-2x-y}Ba_y(PO_4)_3Cl:xCe^{3+},xNa^+$ (x = 0.05, y = 0, 2.45 and 4.90 mole). The results for $[PO_4]^{3-}$ are listed for comparison.

No	Phosphors	V_1	V_2	V_3	V_4	Reference
1	$Sr_{2.45}Ba_{2.45}(PO_4)_3Cl$	941.308	415.376	1040.699	574.428	This work
			432.548			
2	$Sr_{2.9}(PO_4)_3Cl$: $0.05Ce^{3+},0.05Na^+$	949.873	417.834	1028.162	581.508	This work
			456.954			
3	$Sr_{2.45}Ba_{2.45}(PO_4)_3Cl$: $0.05Ce^{3+},0.05Na^+$	943.451	415.376	1036.525	569.701	This work
			432.548			
4	$Ba_{2.9}(PO_4)_3Cl$: $0.05Ce^{3+},0.05Na^+$	934.870	412.917	1026.068	567.335	This work
			427.649			
5	$[PO_4]^{3-}$	938	420	1017	567	Ref. 7

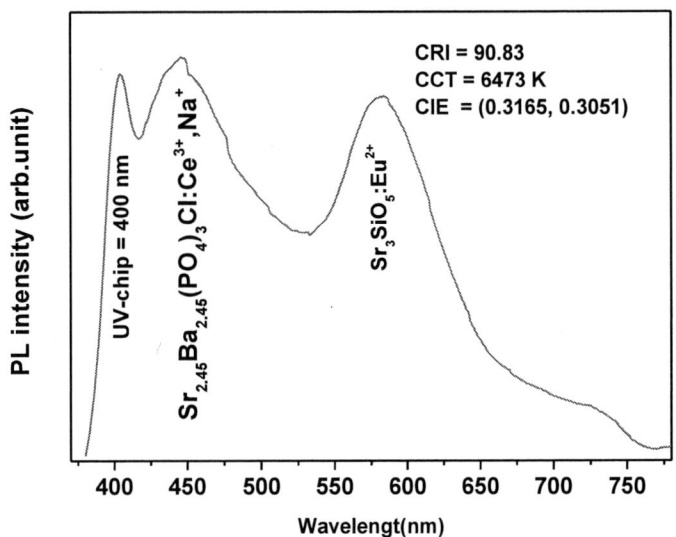

Figure 5.Simulated spectrum of white-LED based on blue emitting $Sr_{2.45}Ba_{2.45}(PO_4)_3Cl:0.05Ce^{3+},0.05Na^+$ and yellow emitting $Sr_3SiO_5:Eu^{2+}$ phosphors

References

1. V. Bachmann, C. Ronda, and A. Meijerink, Chem. Mater. 21,2077(2009).

2. H.S. Jang, Y.H. Won, and D.Y. Jeon, Appl. Phys. B: Lasers Opt. 95, 715(2009)

3. G. Deressa, K.W. Park, H.S. Jeong, S.G. Lim, H.J. Kim, Y.S. Jeong, and J.S. Kim, J. Lumin. 161, 347(2015).

4. G. Deressa, K.W. Park, and J.S. Kim,Chem. Phys. Lett. 645, 42(2016).

5. G. Ju, Y. Hu, L. Chen, and X. Wang, J. Appl. Phys. 111, 113508, (2012).

6. Q. Zeng, H. Liang, G. Zhang, M.D. Birowosuto, Z. Tian, H Lin, Y. Fu, P. Dorenbos, and Q. Su.,J. Phys.: Condens. Matter. 18, 9549(2006).

7. E.J. Baran, J. Raman Spectrosc. 21 391, (1996).

Author Index

Alden, D.	31	Irving, D.	31
Ancona, M. G.	3		
Anderson, T. J.	3	Jang, S.	23
Baik, K. H.	23	Kim, J.	23
Bharadwaj, A.	67	Knorr, D. B. Jr.	9
Bryan, I.	31	Koehler, A. D.	3
Bryan, Z.	31	Koukitu, A.	31
		Krause, J. R.	41
Callsen, G.	31	Kub, F. J.	3
Chang, E. Y.	19	Kumagai, Y.	31
Chung, R.	9		
Collazo, R.	31	Lazarus, N.	89
		Lee, S.	59
Deressa, G.	97	Li, Y.	83
		Liu, S. C.	19
Eddy, C. R. Jr.	3		
Edwards, N.	67	Mahaboob, I.	89
El-Atab, N.	73	Manley, R. G.	67
Enck, R. W.	9	Mudgal, T.	67
Feygelson, T. I.	3	Nayfeh, A.	73
Gaddy, B.	31	Paine, D. C.	59
Ganesh, P.	67	Park, S. H.	47
Garrett, G. A.	9	Pate, B. B.	3
Ge, D.	83		
		Reed, M. L.	9
Han, J.	47		
Hirschman, K. D.	67	Sampath, A. V.	9
Hite, J. K.	3	Shahedipour-Sandvik, S.	89
Hobart, K. D.	3	Sitar, Z.	31
Hoffmann, A.	31	Stokes, E. B.	41
Huang, C. K.	19		
		Tadjer, M. J.	3

Tompkins, R. P.	89
Wheeler, V. D.	3
Xiong, K.	47
Yuan, G.	47
Zhang, C.	47
Zhang, L.	83
Zhu, X. D.	83